FUNDAMENTALS of
SYSTEMS BIOLOGY
From Synthetic Circuits
to Whole-cell Models

FUNDAMENTALS of SYSTEMS BIOLOGY

From Synthetic Circuits to Whole-cell Models

Markus W. Covert
Stanford University

CRC Press
Taylor & Francis Group
Boca Raton London New York

CRC Press is an imprint of the
Taylor & Francis Group, an **informa** business

CRC Press
Taylor & Francis Group
6000 Broken Sound Parkway NW, Suite 300
Boca Raton, FL 33487-2742

© 2015 by Taylor & Francis Group, LLC
CRC Press is an imprint of Taylor & Francis Group, an Informa business

Printed on acid-free paper
Version Date: 20141023

International Standard Book Number-13: 978-1-4200-8410-8 (Paperback)

Library of Congress Cataloging-in-Publication Data

Covert, Markus, author.
  Fundamentals of systems biology : from synthetic circuits to whole-cell models /
Markus W. Covert.
    p. ; cm.
  Includes bibliographical references and index.
  ISBN 978-1-4200-8410-8 (hardcover : alk. paper)
  I. Title.
  [DNLM: 1. Systems Biology. 2. Genetic Variation. 3. Models, Biological. QU 26.5]

  QH324.2
  572.8--dc23
                                                                    2014023781

Visit the Taylor & Francis Web site at
http://www.taylorandfrancis.com

and the CRC Press Web site at
http://www.crcpress.com

Contents

Preface

OVERVIEW

Let's begin our journey into systems biology by comparing complex diseases to a firing squad. While working as a postdoc at the California Institute of Technology, I had the opportunity to know a talented biologist and a well-read Russian who convinced me to read Tolstoy's *War and Peace*. One thrilling highlight of that masterpiece finds the protagonist, Pierre, in front of a firing squad in Moscow. Tolstoy describes the wild thoughts ringing in Pierre's mind as he faces the guns, wondering how he had come to this point. "Who was it that had really sentenced him to death?" As Pierre looks into the nervous eyes of the young soldiers, he realizes that they are not to blame, for "not one of them had wished to or, evidently, could have done it."

So, who was responsible? In all the chaos of the scene and his imminent death, Pierre has a moment of clarity. He realizes that it was *no one*. Rather (and I have added italics for emphasis), "It was a *system*—a concurrence of circumstances. A *system* of some sort was killing him—Pierre—depriving him of life, of everything, annihilating him."

As I read this passage, I could not help but think about the many patients with complex and untreatable diseases that feel virtually the same way as Pierre. What is it that has really sentenced them to death? When they read the newspapers or watch television, they see headline after headline about "the cancer gene" or "the Alzheimer gene," but if these genes truly exist, then why have cures to these diseases eluded us?

I take my answer from Pierre: Such diseases do not depend on any one gene, but on a "concurrence of circumstances"—a system. A *Science* commentary on complex diseases emphasizes this point:

> The most common diseases are the toughest to crack. Heart disease, cancer, diabetes, psychiatric illness: all of these are "complex" or

"multifactorial" diseases, meaning that they cannot be ascribed to mutations in a single gene or to a single environmental factor. Rather, they arise from the combined action of many genes, environmental factors, and risk-conferring behaviors. One of the greatest challenges facing biomedical researchers today is to sort out how these contributing factors interact in a way that translates into effective strategies for disease diagnosis, prevention, and therapy. (From Kiberstis, P. and Roberts, L. *Science.* 2002, **296**(5568): 685. Reprinted with permission from AAAS.)

Put simply, our ability to tackle complex diseases is limited by our ability to understand biological systems. We need ways to explain organismal behaviors in terms of cellular components and their interactions. Even a small number of components can interact in nonintuitive ways; thus, systems-level research requires mathematical and computational strategies.

We are at the cusp of a revolution in understanding the systems-level mechanisms that underlie human disease. However, before the revolution can achieve its full potential, we first must grasp the systems-level behavior of molecules, pathways, and single cells. Vast progress has been made toward this goal in the past 30 years, but much exciting research remains to be completed before the power of these systems-level investigations can be turned to the elucidation and eradication of human disease.

This book seeks to empower you, the student, by aiding you to develop the tools, techniques, and mindset to directly engage in primary research yourself. Whether you are interested in microbes, organs, whole organisms, diseases, synthetic biology, or just about any field that investigates living systems, the intuition that you will develop through the examples and problems in this book will critically contribute to your success at asking and answering important scientific questions.

This book focuses on the use of computational approaches to model, simulate, and thereby better understand complex molecular and cellular systems, research that is often called "systems biology." This field has grown rapidly: There is now an Institute for Systems Biology in Seattle, a new Department of Systems Biology in various institutions (notably Harvard Medical School and Stanford University), a Nature/EMBO (European Molecular Biology Organization) journal, *Molecular Systems Biology*, and an International Conference on Systems Biology that draws over 1,000 people each year. The field has drawn together researchers from nearly every scientific domain. For example, students from biology, bioengineering, computer science, chemistry, biomedical informatics, chemical and systems biology, aeronautical engineering, chemical engineering, biophysics, electrical engineering, and physics have all participated in the systems biology class that I teach at Stanford University.

I teach the material in this book to advanced undergraduate and beginning graduate students over 10 weeks, with two 90-minute lecture sessions per week. Both my class and this textbook are divided in half. The first half, "Building Intuition," focuses on learning the computational tools that underlie systems biology using a simple autoregulatory feedback element as the subject of study. Without grasping these fundamental concepts, you will have great difficulty in forming intuition about the systems-level behavior of molecules and cells, intuition that will strongly contribute to the effectiveness of your systems biology research. The concept of biological feedback and the circuit itself are introduced in Chapter 1, which is followed by five chapters that describe various methods to model this circuit's behavior. Chapter 2 concerns Boolean logic models, Chapters 3–5 focus on ordinary differential equation-based solutions (using analytical, graphical, and numerical methods, respectively), and Chapter 6 describes how to perform stochastic simulations.

The second half of the text, "From Circuits to Networks," applies the tool kit developed in Section 1 to study and model three of the most important and interesting biological processes: transcriptional regulation (Chapter 7), signal transduction (Chapter 8), and metabolism (Chapter 9). Finally, Chapter 10 describes the methods for integrating these diverse modeling approaches into integrated hybrid models using the same techniques that recently led to the creation in my lab of a whole-cell model, the first of its kind.

LEARNING FEATURES

Prerequisites

The material in this book represents a substantial convergence of thought, expertise, and training. Although I believe that this interdisciplinary aspect is one of the most exciting aspects of the field, it also presents a teaching challenge because student backgrounds can be so varied. I therefore encourage students who have never gone beyond high school biology either to take a class or to review a book such as *Molecular Biology of the Cell* (Alberts et al., 2007). I recommend that the biologists at least be acquainted with ordinary differential equations before tackling this material. In addition,

I highly recommend that all students have some experience in mathematical programming. Many of the problems within and at the end of each chapter give hints about potentially useful MATLAB® functions, but this help will not be sufficient if the student has never programmed previously. Sample code appears in the chapters and Practice Problems, end-of-chapter Problems, and their solutions (solutions available online).

Learning Objectives

My goal in writing this textbook is to give you a hands-on, behind-the-scenes tour of this exciting field. After working through this book, I expect that you will

- appreciate the importance of studying biological systems as a whole, rather than as isolated parts;
- be proficient with a broad range of modeling methods relevant to systems and computational biology;
- understand several important studies in the field; and
- apply systems biology approaches to your own research someday!

Specific learning objectives for each chapter are listed at the beginning of the chapter.

Practice Problems

Each chapter contains Practice Problems that explore concepts that may be new to you. The solutions to these problems model how to implement the mathematics described in the text (for example, in MATLAB), how to effectively present quantitative results, and how to conceptualize the biological behavior that underlies our measurements, calculations, and simulations.

Sidebars

Chapters 1, 3, and 8 contain sidebars that delve further into the details or background of concepts presented in the text.

Glossary

Within the text, I have highlighted critical terms that may be unfamiliar to systems biology, biology, biochemistry, or engineering novices. The glossary contains definitions of those terms that focus on our implementations in the text.

Chapter Summaries

Each chapter ends with a few paragraphs that summarize the major concepts and methodologies of the chapter. The summaries also highlight the most critical assumptions, equations, and mathematical treatments that you explored in that chapter.

Recommended Reading

Citations to primary research articles, textbooks, websites, and popular-science works that are directly related to each chapter appear at the end of each chapter. I hope that students will "follow the citations" as they formulate their own research interests—and perhaps discover new ones.

Problems

Finally, each chapter ends with a selection of problems that are intended to give you hands-on experience with computational systems biology. You will explore MATLAB code, practice a variety of modeling techniques, and start developing an intuition for biological systems that will provide the foundation for future scientific endeavors. I provide a few hints to help you along (especially hints to help you navigate MATLAB). The solutions to these problems are provided in a separate solutions manual that will be available to instructors.

RECOMMENDED READING

Alberts, B., Johnson, A., Lewis, J., Raff, M., Roberts, K., and Walter, P. *Molecular Biology of the Cell* (5th edition). New York: Garland Science, 2007.

Kiberstis, P. and Roberts, L. It's not just the genes. *Science* 2002. **296**(5568): 685.

Moore, H. *MATLAB for Engineers* (3rd edition). Englewood Cliffs, NJ: Prentice Hall, 2011.

Acknowledgments

THIS BOOK WOULD NOT exist if not for the support of colleagues, friends, and family. I would first like to especially thank Tiffany Vora, who managed all aspects of the project and made significant contributions at every level. I would not even want to imagine writing this book without her help. The assistance of Mary O'Reilly and Jane Maynard with several of the illustrations made a tremendous difference in terms of the book's look and feel. Michael Slaughter at CRC Press/Taylor & Francis has been amazingly helpful and patient over the years as I worked my way through this material. And, as always, my assistants Jocelyn Hollings, Thi Van Anh Thach, and especially Kimberly Chin were outstanding sources of expert assistance and unwavering support.

Next, I would like to thank my student SWAT team: The teaching assistants for my classes over the last few years who were instrumental in developing practice problems, providing feedback on my lectures, and sharing insights that helped me to understand how to teach this subject in a more intuitive and inspiring way. They are Nathan Barnett, Ryan Bloom, Gautam Dey, Chris Emig, Peiran Gao, Miriam Gutschow, Jake Hughey, Grace Huynh, Anna Luan, Chris Madl, John Mason, Daniel McHugh, Natasha Naik, Jayodita Sanghvi, and Anne Ye. I would also like to thank the students in my classes who provided invaluable feedback on drafts of the manuscript; their comments led to major improvements in the work.

One of the great blessings of my life is to work with an incredible group of people in my lab and department at Stanford University. These outstanding women and men have done so much to educate and inspire me, and I cannot thank them enough. Moreover, I would not be here at this point in my life without my undergraduate, graduate, and postdoctoral mentors—Richard Rowley, Bernhard Palsson, and David Baltimore—and my former labmates in those labs. I am so grateful to have had that time to learn from all of them.

I would like to thank the family that I love with all my heart: first, my parents Michael and Ludwiga, my sisters, Jennifer and Michelle, my brother, Jeff, and all of their children. Next, I thank the family members we added by marriage: David and Connie, Marian and George, Pargie and Will, Betsy, David and Jennie, Heidi and Joe, and Brandon, and all of their children. I am so grateful for the relationships I have with extended family—aunts, uncles, and cousins—and I wish that my grandparents could see how all their incredible work and love made a difference in my life. Finally, to my best friends and children, Peter and Anja, and most of all my truest, deepest friend and confidant, Sarah: here's looking at you, forever.

About the Author

Markus Covert is an Associate Professor of Bioengineering and, by courtesy, Chemical and Systems Biology at Stanford University. He has received the National Institute of Health Director's Pioneer Award and an Allen Distinguished Investigator Award from the Paul Allen Family Foundation. He is best known for the development of the first "whole-cell" computational model of a bacterial cell.

I

Building Intuition

CHAPTER 1

Variations on a Theme of Control

LEARNING OBJECTIVES

- Become familiar with the negative autoregulatory transcription circuit
- Understand the importance of learning different mathematical methods
- Explore common experimental techniques used in systems biology

VARIATIONS

When I was a young piano player, my teacher gave me a new piece to learn that I found fascinating. It was Mozart's variations on a tune that I knew as "Twinkle, Twinkle, Little Star." The piece begins with a simple rendition of the tune and then catapults into a fast series of scales, followed by 11 other variations, including key changes, dramatic shifts in tempo, and plenty else to keep my fingers busy.

What amazed me about this piece was how much I was able to learn by studying a simple melody from almost every technical angle. Mozart took advantage of the tune's ubiquity to explore all kinds of techniques and styles. I was therefore able to focus on learning these techniques without having to relearn a new melody. At the same time, the variations offer a much deeper understanding of the original tune by presenting it simply at first, but then with increasing levels of complexity and nuance. Each variation is a new way of looking at the tune, and as I considered all of the variations together I realized that I would never hear "Twinkle, Twinkle, Little Star" the same way again.

In the next several chapters, I present variations on a ubiquitous theme in biology: The regulation of gene expression by proteins binding to DNA. We examine a relatively simple biological system, going over it again and again with increasingly sophisticated mathematical approaches. My goals are inspired by Mozart: First, I want you to grasp all these approaches without learning new biology, and second, I want you to see this "simple" system in all of its beautiful complexity and nuance. You will find that some of the methods we learn highlight certain aspects of the system, while other methods yield very different insights.

AUTOREGULATION

Our theme falls within the general topic of **control**, which plays a central role in engineered systems. In fact, control systems deeply impact our existence without most people ever being aware of it, which is why the following point by John Doyle, a prominent control theorist at the California Institute of Technology, is useful:

> Without control systems there could be no manufacturing, no vehicles, no computers, no regulated environment—in short, no technology. (Doyle, Francis, and Tannenbaum, 2009, p. 1)

For example, if you look under the hood of your Honda Civic (just a guess) or examine the screen of a Toyota Prius, you are going to find systems that monitor and control virtually every aspect of the car's operation, from engine performance and battery charge to braking and airbag deployment.

Yet control does not just permeate our daily lives: It is the foundation for life itself. Without control systems in biology, our cells could not maintain homeostasis, divide, or carry out any process that relies on **feedback**, control, or "memory" of earlier physical, chemical, biological, or environmental states.

Even in some of the least-complicated organisms, biological control appears to be extensive. You are probably familiar with the so-called **central dogma** of molecular biology, scrawled by high school biology teachers on chalkboards all across the world:

$$DNA \rightarrow RNA \rightarrow Protein$$

in which DNA, filled with information units called genes, can be transcribed to make molecules called messenger RNA (mRNA), which can further be translated to create proteins, molecules that perform many vital cellular functions (of course, there are many caveats to this simplified scheme, many of which we touch on further in this book).

SIDEBAR 1.1   ON NOTATION

For better or worse, each of the organisms that biologists like to focus on—the **model organisms**—is associated with its own, sometimes unique, nomenclature and notation systems. *Escherichia coli* (genus and species names always appear in italics) has a long history of study, and thus its notation system is fairly well standardized. Gene names in *E. coli* (we can shorten the genus name once it is defined) are italicized and usually lowercase (*crp, arcA*), while protein names appear in roman font with a capital first letter. Sometimes, these gene names are actually abbreviations of a description of what the protein does; for example, *arcA* stands for "aerobic respiration control, gene A." Careful notation is part of the accurate and specific language of science and helps us avoid confusing one molecule for another—or writing unnecessarily long sentences.

We start by considering how cells regulate the transcription of DNA into RNA. A cell's ability to create active, operational proteins from DNA-encoded information at the right time and under the appropriate conditions is absolutely critical to cellular survival and, as a result, is carefully and extensively regulated by the cell by what is called a **transcriptional regulatory network**. Figure 1.1 is a graphic depiction of the known transcriptional regulatory network in *Escherichia coli*, a gut bacterium that is probably the best studied of all organisms. To give you an example of how extensive the control is, only about half of the roughly 4,400 genes in *E. coli* are expressed under typical laboratory growth conditions. The network in Figure 1.1 contains most of what we know about *E. coli* transcriptional regulation; the picture includes 116 proteins that regulate transcription (often called **transcription factors**) and 577 **target genes**.

One-half of the transcription factors in Figure 1.1 (58 proteins) appear in red. These transcription factors regulate their own expression; they are transcription factors, but also target genes! This phenomenon is called **autoregulation**. I hope you obtain a sense of how important autoregulation is to the cell simply by seeing how pervasive it is in Figure 1.1, but scientists have also quantified this extensiveness by comparing the *E. coli* network to randomly generated networks of the same size. Specifically, they

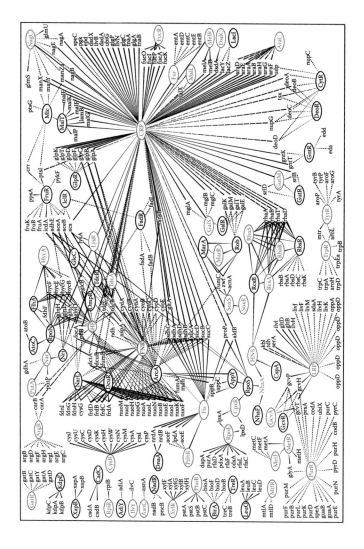

FIGURE 1.1 The transcription regulatory network of the bacterium *Escherichia coli* exhibits extensive autoregulation. Protein factors (ovals) bind DNA and modulate the transcription of target genes. Links between transcription factors and target genes are indicated by connecting lines; solid lines denote activation, dotted lines represent repression, and dashed lines indicate both activation and repression. Note the extensive presence of autoregulation (red), in which a protein activates or represses transcription of its own gene. (Modified from Herrgård, M. J., Covert, M. W., Palsson, B. Ø. *Genome Research.* 2003, 13(11):2423–2434, with permission from Cold Spring Harbor Laboratory Press.)

calculated how many cases of autoregulation one would obtain by chance if the links between transcription factors and target genes were shuffled randomly throughout the network; they found that autoregulation is 30–60 times less likely to occur in a random network that is similar in size to the *E. coli* network (*An Introduction to Systems Biology: Design Principles of Biological Circuits*, which is listed in the Recommended Reading section for this chapter, considers this calculation in more detail). Interestingly, autoregulation is even more common for transcription factors that regulate the expression of large numbers of genes. Approximately 70% of these transcription factors are autoregulated (Figure 1.1), suggesting that autoregulation plays a fundamental role in these cases.

Let's focus on one example of autoregulation, the transcription factor FNR (*fumarate and nitrate reduction, near the center of Figure 1.1). This transcription factor's activity depends on the presence of molecular oxygen, and it regulates *E. coli*'s transitions between aerobic and anaerobic environments: for example, between the gut and fecal environments. Figure 1.2 shows how the FNR transcription factor regulates its own

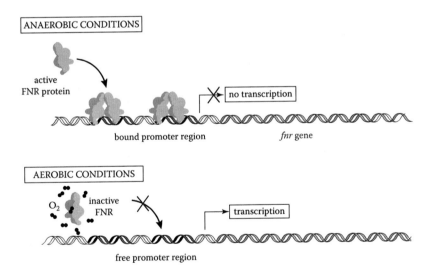

FIGURE 1.2 The *fnr* promoter contains binding sites (dark regions) for its own protein product. In this autoregulatory circuit, pairs of FNR protein molecules, known as dimers, form in the absence of oxygen and are able to bind upstream of the *fnr* promoter (top), preventing RNA polymerase from transcribing mRNA from *fnr*. When oxygen is present (bottom), FNR dimers cannot form, and FNR can no longer repress its own transcription. The *fnr* promoter also contains binding sites for other binding proteins, which are not depicted here.

expression as a **dimer**. The figure depicts a particular location on *E. coli*'s chromosome (a "locus"). The hinged arrow represents the **promoter** (where transcription starts) for the gene that is being expressed (*fnr*), and the helix to the right of the promoter is the part of the gene that actually contains information for making the FNR protein. In the upstream DNA (to the left of the arrow), the dark regions are places where the FNR protein can bind to the promoter of the *fnr* gene.

Binding of FNR to the promoter physically blocks access to its own gene, preventing *fnr* from being transcribed. Whether FNR can bind to DNA depends on the presence or absence of molecular oxygen. When there is no oxygen in the environment, FNR is active, and it binds DNA. When oxygen is present, it binds the FNR protein, which undergoes a structural change that prevents it from binding DNA. Once FNR is reactivated, it represses its own transcription. Interestingly, this means that when this transcription factor becomes active, expression of its gene can decrease.

Active *E. coli* transcription factors can exert a positive, negative, or dual (positive in some cases, negative in others) influence on gene expression (Figure 1.1), but in regulating expression of their own genes (autoregulation), they strongly favor negative regulation. Roughly 70% of the autoregulatory transcription factors in *E. coli* exert a negative influence on their own gene expression.

OUR THEME: A TYPICAL NEGATIVE AUTOREGULATORY CIRCUIT

Our goal for the next several chapters is to consider a typical negative autoregulatory transcription unit somewhat analogous to FNR. A schematic diagram of how this type of circuit works is shown in Figure 1.3. A protein **complex**, RNA polymerase, binds the promoter region just before ("upstream of") the gene. It then travels along the gene, transcribing it to mRNA in the process. Next, the free mRNA is bound by a **ribosome**, which travels along the mRNA and translates it into a protein, our transcription factor. The word *transcribe* will always be used to describe the transition of information from DNA to mRNA, and the word *translate* will indicate the transition from mRNA to protein.

Figure 1.3 also illustrates how expression of the gene is controlled. The protein is activated by an external **stimulus** and becomes able to bind a specific site in the promoter, called an **operator** or *cis*-regulatory element (the darker region of the helix at the left side of the DNA molecule

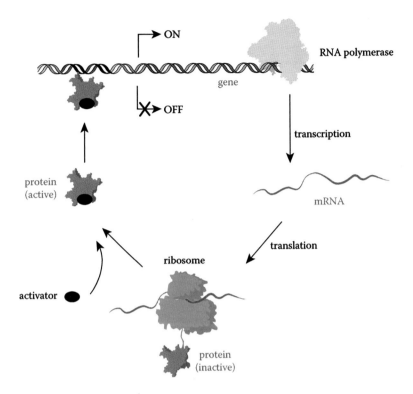

FIGURE 1.3 Our typical negative autoregulatory circuit. Our favorite gene is transcribed into mRNA by RNA polymerase, then translated into protein by the ribosome. The protein is activated by an external signal; the active protein product binds the DNA directly upstream of the gene and blocks access by RNA polymerase. The gene is thus shut off—no more mRNA or protein can be made— and over time, the amounts of protein and mRNA will be reduced.

in Figure 1.3). When the transcription factor is bound to the operator, RNA polymerase is no longer able to bind the promoter region for this gene; as a result, no more mRNA or protein is expressed. After some time, the cellular amounts of both molecules will diminish.

Of course, people have been studying circuits like this for decades, so many experimental and computational approaches have been developed for detailed analyses. Sidebar 1.2 gives you a general flavor of the diversity of approaches currently available. There are also many ways to study these systems computationally. Using **bioinformatics**-based approaches, we could compare sequences of the gene itself or parts of the promoter region to find similar transcription factors in other organisms or genes that are likely to have similar regulation in the same organism, respectively.

Several interesting data-mining strategies and databases have also been developed to help you learn what is already known about any given gene; EcoCyc, which stores information such as that used to construct Figure 1.2, is just such a database.

SIDEBAR 1.2 EXPERIMENTAL MEASURES WE COULD USE TO INTERROGATE OUR CIRCUIT

One of the great revolutions in modern biology came about through the recent development of techniques to make hundreds, thousands, and even millions of measurements simultaneously. We refer to these methods as "high-throughput" or "global" approaches, in contrast to the classical "local" approaches first employed by biologists. It is important to remember that even though these modern approaches appear more powerful, they are also more expensive and need advanced analysis techniques incorporating computer programming and statistics—so if you only want to measure the abundances of a handful of molecules, it may be most efficient to consider classical approaches.

In systems biology, we commonly measure the abundances of RNAs and proteins. First, let us consider how we could measure RNA abundance. If you only wanted to measure the abundances of a few different RNAs, you could probably use a technique such as **northern blotting** or fluorescence in situ **hybridization**, both of which "count" RNA molecules by binding them to other molecules carrying a detectable label. Measuring a fairly large set of RNAs could be achieved with **quantitative polymerase chain reaction (qPCR)**, a highly sensitive but gene-specific technique. Finally, if you wanted to count many thousands of RNAs—or all of the RNAs encoded by an organism—you could take advantage of gene **microarray** analysis or **sequencing** the RNA molecules themselves.

Protein abundance is also a critical component of cellular behavior. To measure the abundance of a small number of proteins, you could use **western blotting**, a technique involving **antibody** detection that is conceptually similar to northern blotting. **Mass spectrometry** would provide measurements of a larger set of proteins and is especially useful for cases in which you do not have an antibody against your proteins. Two-dimensional protein **gel electrophoresis** can provide simultaneous relative measurements of a large variety of proteins and may even capture an organism's entire protein repertoire.

We also have tools to investigate the special case of **DNA-protein interactions**, which form an input to the circuits we consider in this book. The most popular way to detect these interactions is first to cross-link the proteins to the DNA with formaldehyde, then gather either a particular set of DNA-protein interactions with an antibody (chromatin **immunoprecipitation**, or ChIP) or the entire repertoire of DNA-protein interactions from an organism, depending on your needs. If you expect to find, or are interested in, only a small number of binding sites, you would use qPCR of the DNA bound to the protein, but if you wanted to query many binding sites or the entire genome, you could hybridize the pool of DNA that was bound to your protein to a microarray (ChIP-chip) or sequence it (ChIP-seq).

Learn to love this autoregulatory circuit now! You're going to be seeing a lot of it, because how we model the integrated function of this circuit will be broadly applicable to systems biology as a whole. We will use five different approaches on the circuit. First, we will use **Boolean** logic. Next, we will consider sets of **ordinary differential equations**, using **analytical solving techniques, graphical solving techniques,** and **numerical solving techniques** to solve these equations. Finally, we will learn how to build **stochastic simulations.** These methods will form the foundation of your systems biology tool kit.

On to Boolean!

CHAPTER SUMMARY

The goals of the first six chapters of this book are to explore several different modeling techniques and to investigate how they can be applied to molecular systems in cells. We will approach these goals by taking a simple and important biological circuit, the negative autoregulatory circuit, and applying our techniques to analyze it. This circuit is ubiquitous in *E. coli*, and although it is relatively simple in biological terms, you will see that it can behave in ways that are complex.

RECOMMENDED READING

Alon, U. *An Introduction to Systems Biology: Design Principles of Biological Circuits.* Boca Raton, FL: Chapman & Hall/CRC Press, 2007.

Doyle, J. C., Francis, B. A., and Tannenbaum, A. R. *Feedback Control Theory.* Mineola, NY: Dover, 2009.

Neidhardt, F. C., Ingraham, J. L., and Schaechter, M. *Physiology of the Bacterial Cell: A Molecular Approach.* Sunderland, MA: Sinauer Associates, 1990.

EcoCyc: Encyclopedia of *Esherichia coli* K-12 Genes and Metabolism. Home page. http://www.ecocyc.org.

Herrgård, M. J., Covert, M. W., and Palsson, B. Ø. Reconciling gene expression data with known genome-scale regulatory network structures. *Genome Research* 2003, 13(11):2423–2434.

CHAPTER 2

Variation: Boolean Representations

LEARNING OBJECTIVES

- Construct Boolean logic statements to describe a biological system
- Calculate a state matrix, identifying stable states
- Determine dynamics of the system based on state matrix outputs
- Understand how key parameter values affect system dynamics

We start our exploration of the negatively autoregulated circuit described in Chapter 1 using a classical Boolean approach that relies on work performed by Rene Thomas at the Universite Libre de Bruxelles in Belgium. When regulation of gene expression was first discovered, Thomas started exploring how to model it; he began with this approach. It's a perfect way to start building intuition!

BOOLEAN LOGIC AND RULES

The idea behind the Boolean approach is that you identify a number of features of the system that can be represented by one of two states. For example, mRNA might be considered present in a cell or not, and a protein might be considered active or not. In many cases, this Boolean representation is a dramatic simplification; in reality, proteins can have intermediate levels of activity, and an mRNA can be present at an intermediate level. However, the beautiful thing about using a Boolean approach to model this system is that you can easily look at every single state of the system, given any inputs and initial conditions. Such a comprehensive view would be impossible with any of the other approaches considered in this book.

The values that these variables hold are determined by logic rules. For example, an mRNA may be considered present only if the corresponding gene exists in that system or organism, and the corresponding rule might read:

$$mRNA = IF \text{ (Gene)}$$

Figure 2.1 is a drawing of our typical negative autoregulatory circuit from Figure 1.3, represented in a way that will facilitate Boolean logic modeling. Please note that initial uppercase letters denote variables (such as Protein; for simplicity, we continue to refer to mRNA-associated variables as "mRNA") and lowercase letters to denote the physical concepts associated with those variables (mRNA and protein). For Boolean analyses, we do not italicize these variables. This extra notation is intended to help you keep all our terms straight.

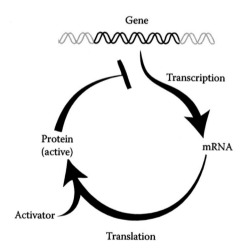

FIGURE 2.1 A Boolean model of our typical negative autoregulatory circuit. Just as we saw in Figure 1.3, our system encapsulates a gene, transcription into mRNA, translation into protein, activation of the protein by an external factor, and transcriptional repression (blunt arrow) following DNA-protein interaction. For simplicity, this Boolean model represents translation and activation as a single process called "translation".

Just as before, a gene is transcribed to mRNA and then translated into a protein that is activated in the presence of an activator. Imagine, for example, a small molecule such as oxygen that binds to the inactive protein to activate it (Figure 2.1 for simplicity lumps the translation and activation processes into one process called "translation"). The active protein then represses further transcription of the gene.

Several variables that describe the system are now defined. There are entities, such as the Gene, Activator, mRNA, and Protein, and processes, such as Transcription and Translation. Each of these variables can have a value of either 0 (absent, off, inactive) or 1 (present, on, active), and our model will keep track of all these values. Furthermore, the presence of the gene and activator can be manipulated fairly easily experimentally by knocking out the gene or adding/removing the activator from the cell's environment. We can therefore consider them as inputs and set their values, as opposed to writing out a Boolean rule.

The Boolean rules for the processes in Figure 2.1 can be written as follows:

$$\text{Transcription} = \text{IF (Gene) AND NOT (Protein)}$$

$$\text{Translation} = \text{IF (mRNA) AND (Activator)}$$

In the transcription rule, transcription occurs if you have a gene and you do not have a protein; that is a repression-based system. Translation takes place if there is both an mRNA and an activator. Importantly, the Boolean value of 0 or 1 does not necessarily have to correspond to a zero **concentration** value: for instance, in cases requiring a significantly high concentration of protein for effective regulation.

The Transcription and Translation rules are instantaneous, meaning that evaluation of their output depends only on the state of the other variables in the system at the same instant in time. If the mRNA and activator are sufficiently present such that their model values are each 1, then Translation is initiated and also has a value of 1.

In contrast, the Boolean rules for the final two entities are somewhat more interesting:

$$\text{mRNA} = \text{IF (Transcription) AFTER SOME TIME}$$

$$\text{Protein} = \text{IF (Translation) AFTER SOME TIME}$$

The values of mRNA and Protein are simply equal to the values for the processes that produce them—but with a time delay. This simple innovation has the nice feature of incorporating a **dynamical** component to the model. The Transcription and Translation variables may be thought of as adding a "memory" of earlier times to the system.

In addition, the time delay creates stable and unstable states in our system. For example, consider Figure 2.2, in which the value of Transcription is changed from 0 to 1 at a certain time point. What follows will be a time period during which Transcription is equal to 1, but mRNA is still equal to 0. We can think of the system's state during this period as unstable because the values of mRNA and Transcription are different from each other, even though our Boolean rule for mRNA suggests that they should be equal.

After this time period has passed, mRNA will be equal to 1. This time period, in biological terms, is the time that it takes for RNA polymerase to locate and transcribe the corresponding gene into mRNA. At some other time point, transcription will be turned off, leading to a change in Transcription from 1 to 0. Again, there will be a certain time period required for the mRNA to be depleted from the system; in this case, the delay is associated with decay of the mRNA molecule. We can make similar statements about the relationship between Translation and Protein.

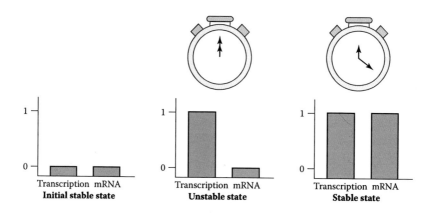

FIGURE 2.2   Stable and unstable states in the Boolean framework. The system is originally in a stable state, where the values of Transcription and mRNA are equal. At time = 0 (shown schematically with a stopwatch), the value of Transcription increases to 1. This state is unstable because the mRNA value is not yet equal to the Transcription value; however, after some time, the mRNA value increases to 1, and the result is a stable state.

STATE MATRICES

As noted, the most compelling thing about the Boolean approach is the capacity to consider *all* states at once. You can do this by drawing a **state matrix**. As mentioned, we have two inputs, Gene and Activator. We also have two processes, Transcription and Translation, which lead to changes in the values of mRNA and Protein, respectively. Since these last two terms correspond to physical entities, it is convenient to define initial conditions with them.

Our state matrix is a chart that holds the values of all of these variables. Let's begin to draw one for this system for the trivial case in which neither the gene nor the activator is present. Intuition would tell you that not much can happen to the circuit without a gene to transcribe, so it is a good example to become familiar with the approach. In this case, the values of our inputs are both equal to 0.

Taking mRNA and Protein as defining the initial conditions, there are four possible starting points for the system: Either mRNA or protein molecules are present individually, or both together, or neither. For each of these starting points and the specific input values, we can calculate what should happen to Transcription and Translation. For example, if neither mRNA nor protein is present, then

$$\text{Transcription} = \text{IF (Gene) AND NOT (Protein)} = 0$$

$$\text{Translation} = \text{IF (mRNA) AND (Activator)} = 0$$

Figure 2.3 shows this part of the state diagram. Notice that the inputs are shown on the top, the mRNA and Protein values are shown at the left, and the values for Transcription and Translation are shown at the lower right.

You might think that this case is not interesting because nothing is happening, but just because nothing is happening does not make a system uninteresting (just look at some of those legendary pitchers' duels in baseball). In this case, there is one interesting feature that I would like you to notice: The system is stable under these conditions. As discussed previously, when the values of mRNA and Protein are equal to the values of Transcription and Translation, respectively, then the system is stable—and unless the system is perturbed in some way, we do not expect any change to occur.

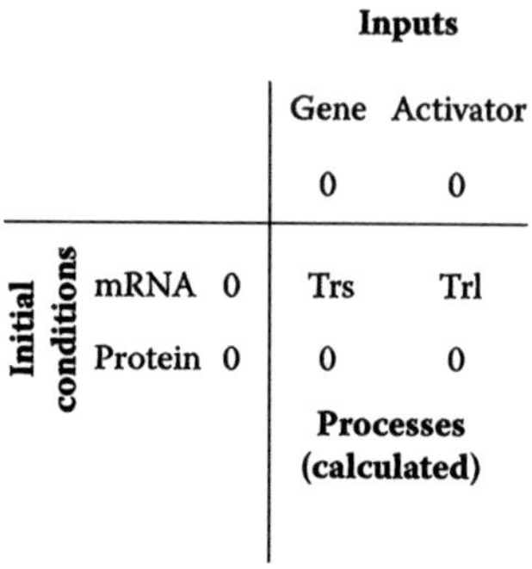

Inputs

		Gene	Activator
		0	0
Initial conditions	mRNA 0	Trs	Trl
	Protein 0	0	0
		Processes (calculated)	

FIGURE 2.3 The state matrix for our negative autoregulatory circuit when neither gene nor protein is present in the system. The values for the variables Gene, Activator, Transcription (Trs), Translation (Trl), Protein, and mRNA are shown. This system is stable under the conditions depicted in the matrix.

STATE TRANSITIONS

Let's see what happens when we change the initial conditions. For example, what happens if we start with some protein? In this case, there still is no gene, activator, or mRNA, so Transcription and Translation are evaluated just as before. This makes sense biologically because there is no gene from which to transcribe.

$$\text{Transcription} = \text{IF (Gene) AND NOT (Protein)} = 0$$

$$\text{Translation} = \text{IF (mRNA) AND (Activator)} = 0$$

The corresponding row of the state matrix can then be added as shown in Figure 2.4. Notice that in this case, the system is no longer stable because the value for Translation (0) does not equal the value for Protein (1). As a result, after some time—when the existing Protein has been depleted from the system—the Protein value will become 0 and the system will once more be stable. This transition is depicted as a red arrow in Figure 2.4, and the stable state is highlighted with a red box.

If you continue with the rest of the initial conditions for the (Gene = 0, Activator = 0) input, you will obtain a full column of the matrix, as shown in Figure 2.5. You can see that there is only one stable state for this system. If the gene and activator are not present, eventually this system will move toward and remain in a stable state in which no mRNA or protein is present and transcription and translation do not occur.

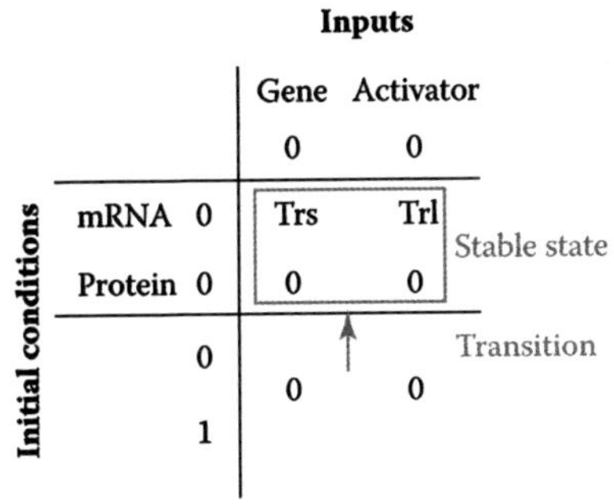

Inputs

		Gene	Activator	
		0	0	
mRNA 0		Trs	Trl	Stable state
Protein 0		0	0	
0			↑	Transition
		0	0	
1				

Initial conditions

FIGURE 2.4 An expansion of the initial state matrix of our typical negative autoregulatory circuit. This expansion depicts the change in the system when the initial conditions include the presence of protein. The updated step appears at the bottom of the figure, and the transition and stable state are highlighted by the red arrow and red box, respectively. Abbreviations are as in Figure 2.3.

Inputs

	Gene	Activator
	0	0
mRNA 0	Trs	Trl
Protein 0	0	0
0	↑	
	0	0
1		
1		
	0	0
1		
1		
	0	0
0		

Initial conditions

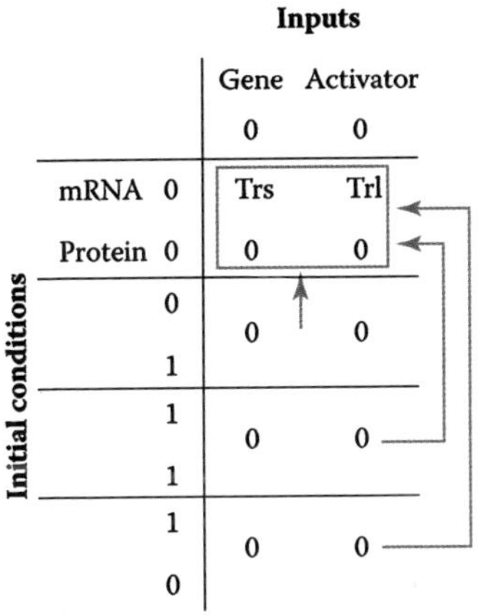

FIGURE 2.5 Full expansion of the initial state matrix of our typical negative autoregulatory circuit. This state matrix has now been fully expanded for the input (Gene = 0, Activator = 0). Only one stable state occurs for this system. Notations and abbreviations are as in Figure 2.3.

Notably, there is no more than a single transition from any set of initial conditions to a stable state.

DYNAMICS

Let's look at what happens when the value of Activator is changed to 1, a case that is shown in Figure 2.6. The value of Gene, and correspondingly the value of Transcription—independent of initial conditions—remains at 0.

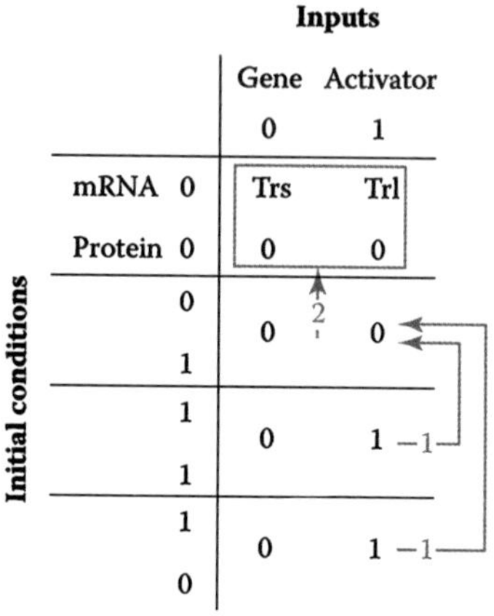

Inputs

		Gene	Activator
		0	1
mRNA 0		Trs	Trl
Protein 0		0	0
0		0	0
1			
1		0	1 −1
1			
1		0	1 −1
0			

(Initial conditions)

FIGURE 2.6 The state matrix for our typical negative autoregulatory circuit when (Gene = 0, Activator = 1). Our circuit has only one stable state possible, regardless of the presence of activator. However, the dynamics differ from the case in which activator is absent (Figure 2.5). Notations and abbreviations are as in Figure 2.3.

Translation depends on the presence of both activator and mRNA, so its value will be 1 for the two initial conditions in which mRNA has a value of 1. As before, you can identify the stable states in this column by looking for rows in which the values of Transcription and Translation are equal to the values of mRNA and Protein. There is one stable state, the same one that we saw in Figure 2.5 in which transcription and translation do not occur and as a result there is no mRNA or protein in the system.

The stable state is therefore identical regardless of the value of Activator (when Gene = 0), but the dynamics are not. Specifically, if the system begins with mRNA, an intermediate state is attained in which protein is being produced. This first transition is labeled with a (1) in Figure 2.6. After some time, however, the protein and mRNA are depleted from the system, and because there is no gene and no transcription, these molecules cannot be replaced. The system then undergoes a second transition, labeled with a (2) in Figure 2.6, to the stable state.

You can visualize these transitions more directly by plotting the values of the processes, as well as mRNA and Protein, over time. The state matrix in Figure 2.6 can be used to create the dynamics plots in Figure 2.7. Although we do not define the length of time explicitly on the x axis (not yet anyway), you can get a sense of the relative timing of events and how the system is behaving.

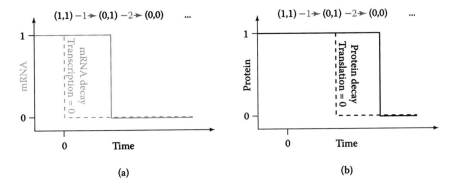

FIGURE 2.7 Dynamic plots of the system, with initial conditions of (mRNA = Protein = 1, Gene = 0). The arrows and numbers (mRNA, Protein) above are for direct comparison with the corresponding trajectory in Figure 2.6. (a) At time = 0, Transcription is set to 0, and after some time, the mRNA decays from the system. (b) This decay causes the value of Translation to be set to 0, and after some time, the protein is lost as well.

PRACTICE PROBLEM 2.1

Using the approach you've just learned, sketch out the column of the state diagram corresponding to a present gene, but no activator. Find the stable states, map out the trajectories, and interpret your results biologically. Finally, draw a dynamic plot of the system's behavior with initial conditions of mRNA = 0, Protein = 1.

SOLUTION

See Figure 2.8 for the column of the state matrix and the dynamics plot. The value of Translation is always 0 because there is no activator, but because there is now a gene, the value of Transcription is 1 when no active protein is present.

There is still only one stable state, but it has changed! Now, all of the initial conditions will lead to a state in which there is always some mRNA but no active protein. We interpret this biologically to mean that no active protein is produced and as a result gene transcription is not blocked, so there will always be some mRNA production (as well as production of the inactive protein, not included explicitly in our model).

Now let's try the most exciting case.

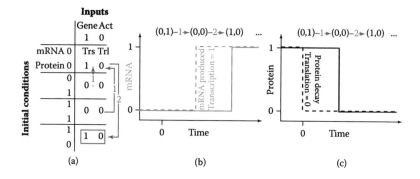

FIGURE 2.8 Solution to Practice Problem 2.1: the state matrix and dynamic plot for our typical negative autoregulatory circuit when (Gene = 1, Activator = 0). (a) This state matrix shows that the stable state of our system has changed compared to the stable states in Figures 2.3–2.6. Notations and abbreviations are as in Figure 2.3. (b), (c) Dynamics plots of system behavior with initial conditions of (mRNA = 0, Protein = 1).

PRACTICE PROBLEM 2.2

Beginning with the stable state you calculated in Practice Problem 2.1, determine what happens to the system when you add an activator to the medium. Calculate the new stable state, if any, map out the trajectories, and explain your results biologically.

SOLUTION

See Figure 2.9 for a matrix that contains all of the states, including the solution to the problem. The system begins in the stable state in which transcription is active and produces mRNA, but no active protein is present.

You will see from the figure that the system has no stable state. Instead, when we add activator to the system, the value of the Translation process is evaluated as 1, so after some time, active protein is produced (Transition 1 in Figure 2.9). Once protein exists, it represses transcription, which then evaluates to 0, leading to depletion of mRNA after some time (Transition 2). Without mRNA, protein can no longer be produced, so after more time, the value of Protein is also reset to 0 (Transition 3). However, the absence of active protein means that transcription can take place again, so after a while, mRNA is produced again (Transition 4).

Bottom line: This system exhibits **oscillation**! It's fascinating that in this reduced model with a highly simplified Boolean representation,

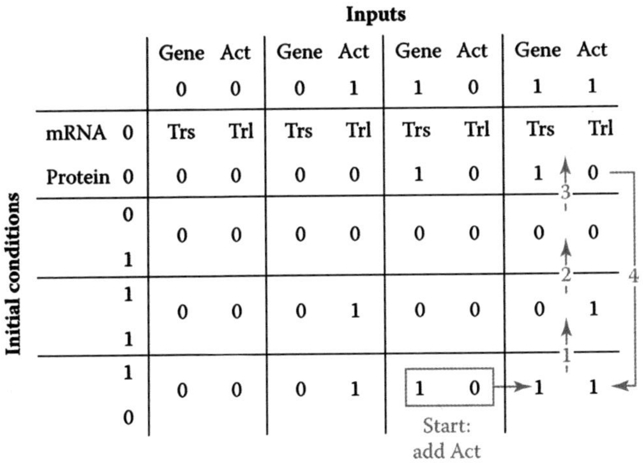

		Inputs							
		Gene Act		Gene Act		Gene Act		Gene Act	
		0 0		0 1		1 0		1 1	
	mRNA 0	Trs	Trl	Trs	Trl	Trs	Trl	Trs	Trl
	Protein 0	0	0	0	0	1	0	1	0
Initial conditions	0 / 1	0	0	0	0	0	0	0	0
	1 / 1	0	0	0	1	0	0	0	1
	1 / 0	0	0	0	1	1	0	1	1

Start: add Act

FIGURE 2.9 The complete state matrix for our typical negative autoregulatory circuit, including the case in which (Gene = 1, Activator = 1). This solution to Practice Problem 2.2 reveals that, under these conditions, our circuit has no stable state. The trajectory specific to (Gene = 1, Activator = 1) is highlighted in red; abbreviations are as in Figure 2.3.

we are nonetheless able to predict some pretty interesting and complicated dynamics. We've been able to push from a stable state (Figures 2.2–2.8) into an unstable state (Figure 2.9). The circuit perpetually oscillates when both gene and activator are present, and all this information can be modeled with just this set of simple Boolean rules.

Another thing to notice here is that Figure 2.9 contains *all* of the possible states of this system. Every combination of input and initial condition values has been evaluated, and all of the results appear on one chart. You can tell where the stable states are, and you know how any combination of factors leads to those stable states—or to the oscillating case of (Gene = 1, Activator = 1).

From the state matrix, it is relatively easy to draw time courses of the different variables as well. For example, Figure 2.10 shows the dynamics of mRNA, protein, transcription, and translation over time for the series of transitions highlighted in Figure 2.9. Again, once the activator is added, Translation is immediately evaluated to 1, followed by Protein transitioning to 1 after some time period. The presence of protein cuts off mRNA production, and mRNA decays after time, halting translation and leading to depletion of protein and subsequent transcription in a new iteration of the cycle.

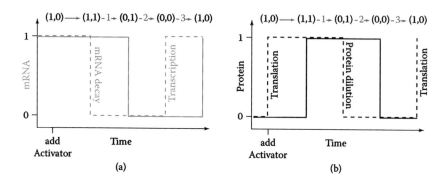

FIGURE 2.10 The dynamics over time for the (a) mRNA and (b) Protein variables from Figure 2.9. The corresponding transitions in the state matrix appear at the top. mRNA decay and protein dilution are inferred in the absence of transcription (gray dashed line) and translation (black dashed line), respectively.

TIMESCALES

One key assumption was made in drawing Figure 2.10: The time periods for mRNA production as well as protein production and depletion were considered to be roughly equal. In reality, the time periods for mRNA and protein production and depletion can all differ significantly from each other.

Let's consider production first. How long does it take to make a typical mRNA? We can use some observations from the literature to make a rough estimate; I've compiled some of these numbers in Table 2.1. A typical RNA polymerase in *Escherichia coli* travels at a rate of nearly 50 nucleotides per second, and a typical gene is approximately 1,000 nucleotides long. Those **parameters** lead us to an estimate of at least 20 s to make the transcript, and if we add in the time required for RNA polymerase to find the promoter, bind, and eventually finish transcription, a good estimate for mRNA production might be 1–2 min.

Protein production turns out to occur on a similar timescale. Ribosomes attach to free mRNAs one after the other in "polysomes" that look like pearls on a string. In this way ~10 proteins can be made simultaneously from a single transcript. The ribosome also travels at ~50 nucleotides (or ~16–17 amino acids produced) per second, which gives us ~20 s for **elongation** of a 333-amino-acid peptide from the 1,000-nucleotide-long gene. Again, when initiation and termination are taken into account, we can say it takes 1–2 min from mRNA existence to protein creation.

TABLE 2.1   **Rough Estimates of Key Parameters Related to Gene Expression in *E. coli***

Number of genes in the *E. coli* genome	~4,400
Average gene length	~1,000 nucleotides
Time required to produce mRNA from a gene	~1–2 min
Time required to produce protein from mRNA	~1–2 min
Half-life of a typical mRNA	~3–8 min
Half-life of a typical protein	~5–10 h
Doubling time during growth in a rich environment	~30 min

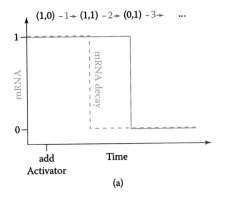 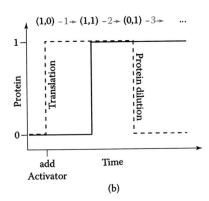

FIGURE 2.11   Adding the timescale information to the dynamics over time shown in Figure 2.10. Notice the loss of cyclical behavior within a few *E. coli* generations. Symbols and notation are as in Figure 2.10.

Now it's time to consider depletion. The mRNA decay rates that have been measured in *E. coli* indicate a typical mRNA **half-life** (the time it takes for one-half of the mRNA to decay) of 3–8 min. The measured protein half-life is substantially longer, probably on the order of several hours on average (although certain proteins are much less stable).

Loss by cell division is called **dilution** and has its own timescale—the time scale of cell growth. For *E. coli* growing under typical lab conditions, we can assume that cell division occurs roughly every 30 min.

The division time is significantly longer than the mRNA decay rate but much shorter than the protein decay rate. Thus, mRNA depletion will be largely driven by decay, with a half-life of ~5 min, while protein depletion will be mostly caused by dilution, with an effective half-life of ~30 min.

Adding what we know about timescales to our time course plots has a dramatic effect on the predicted behavior of the negative-feedback loop (Figure 2.11). We begin at the same point as in Figure 2.10, with mRNA but no protein, and then add the activator. Translation is initiated, and we assume it takes on the order of a few minutes to produce enough protein for our Protein variable to equal 1 (transition 1 in Figure 2.9). Then, transcription is repressed, and mRNA decay leads to loss of mRNA from the system within ~10 min more. Loss of mRNA means that protein will eventually be diluted, but this process is going to take much longer than the buildup of either species or mRNA decay.

As a result, over a short experiment covering a couple of hours, our predicted behavior does not look cyclic at all (although if you plotted for long enough, you would see Protein return to zero and the cycle would continue). In addition, our result suggests that if you want to change the dynamic behavior of the system, a critical parameter will be the protein half-life. In this case, if we reduce the time it takes for protein to be depleted, we return to the oscillatory case of Figure 2.11. People have designed proteins with short half-lives for just this reason, as you will see.

ADVANTAGES AND DISADVANTAGES OF BOOLEAN ANALYSIS

The Boolean approach that we have described here has some advantages and some disadvantages. The disadvantages are fairly obvious: The rules and values are gross simplifications of the underlying biology, so we cannot necessarily rely too much on the results (or predictions) that we obtain. We will have to be more thorough and read the published literature, conduct experiments, or both, especially if we have a particular biological system in mind.

But, the main advantages of the Boolean approach—the method's simplicity, and our ability to consider every single state in a system—are formidable. All you need to perform the analysis described here is paper and pencil to obtain a significant amount of intuition about the system. Furthermore, if not much is known about a system, you can often still guess at Boolean rules and begin to make testable predictions.

In the next chapter, this method is compared with ordinary differential equations.

CHAPTER SUMMARY

To model our autoregulatory circuit, we first consider the Boolean method, wherein all of the variables can only hold values of 0 or 1. Equations are written that describe whether a biological process occurs, given the values of the inputs and the initial conditions of the system. Another set of equations, which gives the output conditions, depends on the processes, but only "after some time." These equations therefore add a dynamic component to the model. All of the inputs, processes, and conditions can be represented together in a state matrix, and the state matrix can be used to draw time plots of the system's dynamics. The main advantages of the Boolean method are that (1) the equations are simple to write and evaluate; and (2) we can see all of the solutions in a single matrix. The disadvantages are that a Boolean representation is often too simplistic, and that without some additional information about relative timescales, the method can lead to incomplete, and possibly even misleading, predictions.

RECOMMENDED READING

Bolouri, H. *Computational Modeling of Gene Regulatory Networks—A Primer.* London: Imperial College Press, 2008.

Gardner, T. S., Cantor, C. R., and Collins, J. J. Construction of a genetic toggle switch in *Escherichia coli. Nature* 2000, **403**(6767): 339–342.

Jacob, F. and Monod, J. On the regulation of gene activity. *Cold Spring Harbor Symposium on Quantitative Biology.* 1961, **26**: 193–211. http://symposium.cshlp.org/content/26/193.full.pdf+html.

Selinger, D. W., Saxena, R. M., Cheung, K. J., Church G. M., and Rosenow, C. Global RNA half-life analysis in *Escherichia coli* reveals positional patterns of transcript degradation. *Genome Research* 2003, **13**(2): 216–223.

Thomas, R. Boolean formalization of genetic control circuits. *Journal of Theoretical Biology* 1973, **42**(3): 563–585.

PROBLEMS

PROBLEM 2.1
Boolean Analysis of a Positive-Feedback Network

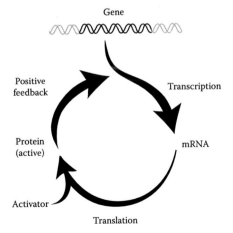

First, let's practice with a system (shown above) that is similar to the system in Figure 2.1, except that this system includes positive feedback instead of negative feedback. Just as before, transcription produces mRNA, translation yields protein, and an activator must be present for translation to occur. The gene and the protein must both be present for transcription to occur. The inputs to this system are the gene and the activator. You are interested in the changes in the mRNA and protein concentrations over time.

a. Write Boolean equations for each of the reactions and mRNA and protein.

b. Construct the state diagram of your system. Circle all of the stable states.

c. Beginning with initial conditions Protein = 1, mRNA = 0, Gene = 1, and Activator = 0, plot the dynamics of the system. Label the changes that occur (for example, the onset of transcription).

d. Imagine that the system starts with mRNA but no protein; both the gene and the activator are present. Describe the progression of the system. How does this progression change if you take the relative timing of protein and mRNA decay into account?

PROBLEM 2.2
Two Interlocking Regulatory Circuits

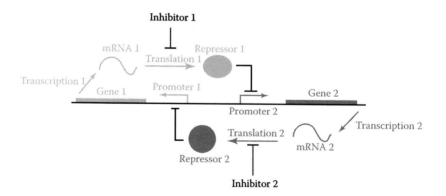

Now, we'll study a regulatory system with two proteins, each of which negatively regulates the expression of the other. In this case, there are also two inhibitors, each of which blocks translation for one of the two mRNA types. A diagram is shown above. Interestingly, this type of circuit, called the *toggle switch*, was the first **synthetic circuit** (see Gardner, Cantor, and Collins, 2000).

To pursue a Boolean analysis of this system, we will assume that the promoters and genes are always present and therefore need not be included in our equations or state matrix. Inhibitor 1 and Inhibitor 2 will be our inputs. There can be no translation of the first repressor if Inhibitor 1 is present and no translation of Repressor 2 if Inhibitor 2 is present. The molecules of interest will be mRNA 1, Repressor 1, mRNA 2, and Repressor 2. The processes we will track are Transcription 1, Translation 1, Transcription 2, and Translation 2.

a. Write Boolean rules for all of the molecules of interest and processes for this system.

b. Calculate the state matrix. Circle or highlight any stable states.

c. Consider the stable state in which Inhibitor 1 is present and Inhibitor 2 is not. At time = 0, the culture medium is suddenly changed such that Inhibitor 2 is now present and Inhibitor 1 is not. Draw a time plot and describe the progression of the system in terms of state transitions.

d. Using some of the typical timescales discussed in this chapter, how long do you think the experiment described in (c) will take to reach a new stable state?

PROBLEM 2.3
The *lac* Operon Regulatory Network Using Boolean Logic

Now, let's apply our Boolean methods to the analysis of a real, and considerably more complicated, bacterial regulatory network. The *lac* **operon** is required for the transport and metabolism of lactose in several enteric bacteria, including *E. coli*. The *lac* operon is still used as a powerful model for investigating aspects of gene regulation and provides the basis for many synthetic gene circuits. For more information on the pioneering work of Francois Jacob and Jacques Monod, who received the 1965 Nobel Prize in Medicine or Physiology for their work on gene regulation, see their 1961 work (See Jacob and Monod, 1961).

The *lac* operon encodes three genes: *lacZ* (encoding the enzyme β-galactosidase, which cleaves lactose into glucose and galactose), *lacY* (the permease that transports lactose into the cell), and *lacA* (a transacetylase). The operon is controlled primarily by the *lac* repressor (LacI), which prevents transcription in the absence of lactose, and the glucose-regulated activator CRP (cyclic adenosine monophosphate receptor protein). You will learn more about CRP in Chapters 7 and 10; for now, it is sufficient to know that CRP induces expression of the *lac* genes and is inactivated by glucose.

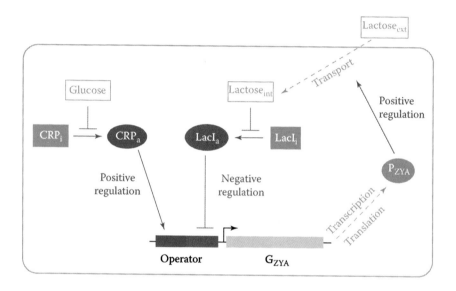

In the simplified version of the *lac* network illustrated above, there is one gene (G_{ZYA}) that is transcribed and translated to produce protein (P_{ZYA}) in the presence of an activator (CRP_a; CRP is active in the absence of glucose) and the absence of a negative regulator ($LacI_a$; LacI is active in the absence of lactose). P_{ZYA} transports lactose into the cell if external lactose ($Lactose_{ext}$) is present.

The two inputs into this system are glucose and external lactose.

The four quantities that we wish to track are CRP_a, $LacI_a$, internal lactose ($Lactose_{int}$), and P_{ZYA}. The four processes that we care about are the transport of external lactose into the cell, the synthesis of P_{ZYA} (we will lump transcription and translation into one process in this model), and the activation of LacI ($LacI_i$ to $LacI_a$) and CRP (CRP_i to CRP_a).

Assume that the inactive regulators, CRP_i and $LacI_i$, are present at stable levels inside the cell, and that G_{ZYA} is always present. These assumptions allow you to remove these quantities from your equations.

a. Initially, assume that activation of CRP and LacI happens much faster than transcription or transport. These assumptions allow you to eliminate two quantities and two processes from the system of equations. Write your equations for this simplified model.

b. Build and fill in your state diagram. You may do this by hand or in Excel or MATLAB (Excel may be more useful in this particular case). Highlight the stable states of the system.

c. Is the system capable of oscillating? Identify any cyclic behavior and explain the logic behind it in biological terms.

d. Assume now that lactose transport across the membrane is slightly leaky. What will happen to the oscillations from (c) when external lactose is present in the absence of glucose? Could this leakiness be a useful attribute of the real system? Explain why or why not in biological terms.

e. Now consider the network without assuming instantaneous activation of CRP and LacI. What is your new set of equations?

f. Build and show your new complete state diagram for the system in (e).

g. Highlight the stable states of the system in (e). Can this new system oscillate? Explain why this network behaves differently from the system in (c) under conditions of abundant external lactose and no glucose. Was the initial assumption to separate the timescales a reasonable one? Justify your answer.

PROBLEM 2.4
Bacterial Chemotaxis

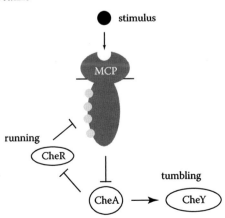

So far, we have focused on the regulation of gene expression, but the Boolean approach can also be used to analyze the dynamics of other biological circuits, such as signaling networks, protein kinase cascades, and metabolic pathways. Here, we analyze a well-studied phenomenon: bacterial chemotaxis. In the lab, bacteria are most often cultured in homogeneous environments that provide all of the nutrients and energy sources that the microorganisms need to grow and **proliferate**. However, in nature, the environment in which the bacteria reside can be heterogeneous and dynamic. Bacteria need to be able to quickly sense the conditions of the surrounding environment and move toward food sources or away from dangerous chemicals or toxins. This process is called "chemotaxis". This video gives a straightforward overview of the process: "Bacterial Chemotaxis in Plain English," http://www.youtube.com/watch?v=1wW2CZz6nM4& feature=related.

We will investigate the dynamics of bacterial motion using Boolean modeling. The above figure is a simplified representation of the chemotaxis network in *E. coli*. The m̲ethyl-accepting c̲hemotaxis p̲rotein (MCP) is a transmembrane protein that binds potential food sources in the environment (stimulus). In the absence of these signals, the CheA protein autophosphorylates itself and transfers this phosphate group to the CheY and CheR proteins. The phosphorylated "active" form of CheY stimulates clockwise motion of the *E. coli* flagellum, which leads to tumbling. In contrast, when CheR is phosphorylated, it is inactive. When MCP is bound to a stimulus, it inhibits the kinase activity of CheA, leaving CheY in the inactive form and CheR in the active form; the bacterium therefore "runs." However, CheR can also methylate the MCP **receptor**, which can cause CheA to regain its phosphorylation activity. This system therefore exhibits a form of negative feedback.

Assume that all of the activation and repression steps happen after some time. Also, assume that CheA and CheR have a default active state (they are activated in the absence of the repressor) and that the repression of MCP by CheR overcomes activation by a stimulus.

We model four processes of this system: activation of MCP (ActMCP), activation of CheA (ActCheA), activation of CheY (ActCheY), and activation of CheR (ActCheR). You are interested in the active forms of MCP, CheA, CheY, and CheR. We have one constant input into the system: the stimulus.

a. Write equations for each of the reactions and molecules of interest.

b. Calculate the state matrix and highlight any stable states.

c. Assume that your system starts from a state in which no stimulus is present and none of the proteins of interest is activated. Trace the dynamics of the system on your state matrix as the system moves from state to state. What happens to the system? What does your observation imply about bacterial chemotaxis in the absence of a stimulus?

d. Starting from the state that you ended with in (c), assume that a stimulus is suddenly added to the environment. What happens to the network? Trace the new dynamics on a state matrix and graph the behavior we would expect the bacterium to exhibit (running or tumbling) over time.

e. Now assume that the bacterium has a mutation in MCP that prevents its negative regulation by CheR, such that

$$ActMCP = IF\ Stimulus$$

Draw a new state matrix corresponding to this new set of rules, determine the stable state of the new system with no stimulus present, and then trace the dynamics of the system on addition of stimulus using a state matrix. How is your answer different from (d)? What is the purpose of CheR's negative regulation of MCP activity?

CHAPTER 3

Variation: Analytical Solutions of Ordinary Differential Equations

LEARNING OBJECTIVES

- Draw compartment diagrams that integrate system inputs and outputs

- Write a set of ordinary differential equations (ODEs) based on compartment diagrams

- Solve simple ODEs analytically

- Compare various systems based on key equation parameters, such as steady-state solution, step response, and response time

This chapter and the next two chapters focus on the workhorse of computational systems biology: sets of ODEs based on **conservation of mass** and **mass action kinetics**. Most models that are published in systems biology papers consist of a set of ODEs.

Steven Strogatz, an outstanding mathematician and teacher at Cornell University, says that every modeler should have three tools for solving sets of ODEs: analytical, graphical, and numerical solving techniques. We are going to investigate all of these tools in the context of our simple circuit. First, we'll solve the ODEs analytically to figure out what they are telling us about what feedback does and why it might be advantageous for a cell to use it.

SYNTHETIC BIOLOGICAL CIRCUITS

We can also begin to look at "real" circuits. The quotation marks around "real" are there because the circuits exist, but not naturally. Instead, they were assembled using the techniques of molecular biology and are called synthetic circuits. Figure 3.1 shows two synthetic gene transcription circuits, one lacking feedback and the other with feedback. The transcription units in both cases involve an activator to control gene expression and a reporter to indicate when transcription is active. The activator, anhydrotetracycline (aTc), binds a negative regulatory protein called TetR and prevents it from binding the promoter region of the gene. The "double negative" of aTc-inhibiting TetR repression leads to a net activation of gene expression when the transcription unit is in the presence of aTc.

The reporter protein, encoded by the gene, is a fluorescent protein from jellyfish called GFP (for green fluorescent protein). When the gene is transcribed and translated, these proteins fluoresce green, and this fluorescence can be detected and recorded.

The transcription unit in Figure 3.1a does not undergo negative autoregulation; it is simply *gfp* with a binding site for TetR in the promoter region. TetR is produced **constitutively** by another gene in this strain of *E. coli*, so the activity of the gene depends on whether sufficient aTc is present to prevent TetR from binding the *gfp* promoter. If aTc is present, gene expression occurs, and we can detect fluorescence; otherwise, there is neither expression nor fluorescence.

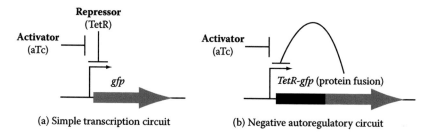

(a) Simple transcription circuit (b) Negative autoregulatory circuit

FIGURE 3.1 Synthetic transcription circuits without (a) or with (b) negative autoregulation. Expression of green fluorescent protein (GFP) is negatively regulated by binding of the repressor TetR, which is either encoded at another locus (a) or fused to the GFP coding sequence for autoregulation (b). TetR is itself inhibited by binding of the activator anhydrotetracycline (aTc), a chemical that can be added to the medium by the investigator. (Reprinted from Rosenfeld, N., Elowitz, M.B., and Alon, U. *Journal of Molecular Biology* 2002, **323**(5): 785–793, with permission from Elsevier.)

In contrast to the simple transcription unit of Figure 3.1a, the circuit drawn in Figure 3.1b includes autoregulatory feedback. In this case, the gene encodes a **fusion protein**: The *gfp* gene follows directly behind the *tetR* gene, so they are transcribed and translated as a single protein. Here, *E. coli* does not produce TetR constitutively (the native *tetR* is absent from this strain), so TetR is only present when gene expression of the synthetic circuit has occurred. The absence of aTc prevents expression of the fusion gene just as before. The presence of aTc, however, induces expression of the circuit and greatly increases the concentration of TetR::GFP (the double colon indicates the fusion protein). TetR binds tightly to aTc, so when enough TetR molecules are present, all aTc molecules are bound and effectively inactivated, allowing TetR to repress further gene expression.

These two transcription units were constructed to experimentally investigate the differences between a system with feedback and one without it. We are going to address the same question here using a set of ODEs that describe the system.

FROM COMPARTMENT MODELS TO ODES

I begin by giving you a brief introduction to ODEs using a rain gutter analogy. When I was young, I once kept measuring cups outside in the rain so that I could determine how much rainfall Palo Alto was receiving in an hour on rainy days. Let's pretend that I wanted to use that rainfall data and other information to determine what was happening to the water in the rain gutter on my house.

Figure 3.2a depicts a house in the rain, complete with rain gutter and drainpipe. As the rain pours into the gutter, water can either flow out of the drainpipe or build up in the gutter. We can indicate this conceptually by using a **compartment model**, as illustrated with a compartment diagram in Figure 3.2b. Here, the rate of rainfall *Rain* and the rate of drainage *Drain* are both considered as functions of time; *GW* indicates the amount of water in the rain gutter.

Let's assume that I can obtain a decent estimate of the rain and drain rates with my measuring cups and a stopwatch, and I want to see what happens to the amount of gutter water over time. I assume that water is not entering the gutter from anywhere but the rain and not leaving from anywhere except the drain. If I know the level of gutter water at a certain time t and want to predict the level after a short time period Δt, I can write the following equation:

$$GW(t + \Delta t) = GW(t) + Rain(t) \cdot \Delta t - Drain(t) \cdot \Delta t \qquad (3.1)$$

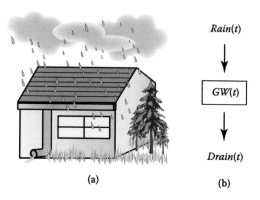

(a) (b)

FIGURE 3.2 Conceptualizing ODEs with a rain gutter analogy. (a) During a rainstorm, water enters the gutter at rate *Rain*, flows out of the drainpipe at rate *Drain*, or builds up in the gutter to an amount *GW*. (b) The compartment diagram for the analogy indicates that water only enters the gutter from the rain and only exits the gutter from the drain; the diagram also captures the dependence on time of these processes.

In other words, the amount of gutter water at the later time is simply equal to how much water was in the gutter before, plus whatever fell in, minus whatever poured out. We calculate the amount of water that fell in by multiplying our gutter rainfall rate, which would be some amount over time, by the time interval Δt. Similarly, we multiply the drain rate by the time interval to determine the amount of water that poured out.

If we rearrange Equation 3.1 by subtracting $GW(t)$ from both sides and dividing by Δt, we obtain:

$$\frac{GW(t+\Delta t)-GW(t)}{\Delta t} = Rain(t)-Drain(t) \qquad (3.2)$$

If we then take the limit of both sides of this equation as $\Delta t \to 0$, we obtain an ODE:

$$\frac{dGW}{dt} = Rain(t)-Drain(t) \qquad (3.3)$$

where dGW/dt is the derivative of the amount of gutter water with respect to time. If I have a good enough understanding of how *Rain* and *Drain* depend on time or on *GW*, I can make some interesting predictions. For example, I can predict the amount of water in the gutter at future times. I can also predict the effect of perturbations (for example, clogging the drain) on the time it takes for the gutter to overflow.

And there we have it—from a grade-school science fair project to an ODE! Systems of ODEs have been used for all kinds of applications. For example, ODEs not much more complicated than our rain gutter example have been used to model the spread of human immunodeficiency virus (HIV) from cell to cell and from person to person, leading to recommendations that had an impact on health policy. They have been used to model waste removal in patients undergoing dialysis and to describe the conversion of sugars to biofuel in microorganisms. Right now we're going to use them to describe our transcription units from Figure 3.1 in an effort to understand the importance of autoregulation.

In all of these cases, the general form of the ODEs looks like a generalized form of Equation 3.3 with more inputs or outputs:

$$\frac{dx}{dt} = \sum Rates_{production} - \sum Rates_{loss} \tag{3.4}$$

In other words, the change in amount or concentration of some entity x over time is equal to the sum of the rates of production of x minus the rates that lead to loss of x.

Take a moment to consider what we have just done. We started with a cartoon-like picture of a system (Figure 3.2a), moved first to a compartment model and diagram (Figure 3.2b), and then continued to a mathematical equation (Equation 3.3) that we can use to analyze, interpret, and even predict the behavior of the system.

If you can learn how to encode a biological network schematic into a set of equations like this, you are well on your way to making substantial, independent contributions to biological research! If your educational background is in biology instead of engineering, you may wonder whether you will be able to catch up with the engineers. Good news: In many ways, biologists are better suited for this part of model construction because it requires strong knowledge and intuition about the system. (In fact, I was once told this by the head of systems biology at a large pharmaceutical company!) If you have that knowledge or intuition, drawing the block diagrams and writing (and even solving, as you will see) the ODEs become fairly straightforward.

Let's practice by writing the equations for our feedback system with a few changes from Chapter 2. First, let's assume that the activator is always present—as you already learned, solutions without activator are somewhat trivial. Second, we will explicitly include loss of the mRNA and protein in this chapter. Remember that loss can occur by degradation of the molecule or by dilution as cells grow and divide.

PRACTICE PROBLEM 3.1

Given the system in Figure 3.3, draw a compartment model for the mRNA and protein concentrations. Write the equations that describe how these concentrations change over time. Decompose the overall loss terms into decay and dilution components.

SOLUTION

The two compartment diagrams appear in Figure 3.4. Notice that there are two compartments now, and that each compartment has two possible ways for either mRNA or protein to be lost.

Using these diagrams, it is relatively straightforward to write the two ODEs for this system:

$$\frac{d[mRNA]}{dt} = Rate_{transcription} - Rate_{decay} - Rate_{dilution} \qquad (3.5)$$

$$\frac{d[Protein]}{dt} = Rate_{translation} - Rate_{decay} - Rate_{dilution} \qquad (3.6)$$

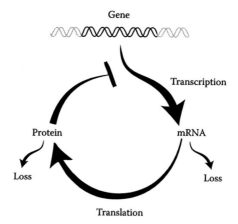

FIGURE 3.3 Slightly altering our feedback system for Practice Problem 3.1. This updated schematic includes the loss of mRNA and protein molecules as well as their production by transcription and translation, respectively. As in the original circuit, the protein product of this updated circuit represses its own transcription.

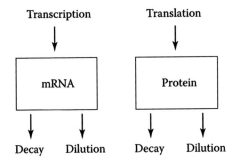

FIGURE 3.4 Compartment diagrams illustrating the mechanisms for creation and loss of mRNA (left) and protein (right) in our updated autoregulatory circuit. All of these processes are associated with specific rates.

SPECIFYING AND SIMPLIFYING ODES WITH ASSUMPTIONS

Now, we've accomplished the first step: writing the equations. However, these rate terms are not specific. How do we add detail to these rate expressions? There are several assumptions we can make about a given rate term. The simplest assumption would be that a given rate is zero or small with respect to other terms in the equation. Removing or simplifying terms as appropriate can make equations much easier to solve.

For example, you learned in Chapter 2 that protein loss by decay often takes much longer (occurs at a much slower rate) than loss by dilution. As a result, the decay rate will be substantially lower than the dilution rate, and we could write that $Rate_{decay} + Rate_{dilution} \approx Rate_{dilution}$, or $Rate_{decay} \approx 0$. In contrast, mRNA loss by decay is typically faster than the dilution rate, so we make the opposite assumption: $Rate_{decay} + Rate_{dilution} \approx Rate_{decay}$, or $Rate_{dilution} \approx 0$.

A second, related assumption is that a rate for a given system does not change significantly under the conditions that interest us. For example, certain genes are known as **housekeeping genes** because their protein products are so critical to cellular function that they are essentially always transcribed at a nearly constant rate. These genes are often used as controls in experiments monitoring gene or protein expression. For housekeeping genes, we would replace $Rate_{transcription}$ with a constant, such as k_{trs}, with units of concentration over time. This constant is a parameter of the model, which needs to be assigned a value to solve numerically. Ideally, we would measure k_{trs} for our circuit; if that were not possible, we would need to estimate it.

Another common assumption is that the process follows mass action kinetics: that the reaction rate is proportional to the product of the reactant concentrations. For example, mRNA decay depends on a reaction between the mRNA molecule and a class of enzymes called ribonucleases (RNases). Assuming that the rate follows mass action kinetics, the rate of mRNA decay at a given moment in time is proportional to the product of the mRNA and RNase concentrations at that time. We then write:

$$Rate_{decay} \propto [mRNA][RNase] \qquad (3.7)$$

Often, this kind of equation is simplified by assuming that certain reactants such as RNase are present at constant levels. If we simplify Equation 3.7 in this way and define a kinetic constant to describe the proportionality between the mRNA concentration and the transcription rate, we obtain:

$$Rate_{decay} = k_{mdec}[mRNA] \qquad (3.8)$$

In other words, the more mRNA that exists, the more that will be degraded in a given interval of time. Moreover, the units of k_{mdec}, 1/time, are different from the translation constant k_{trs}, which has units of concentration over time. Looking at Equation 3.8, you may be asking: "Wait a minute, where did the RNase concentration go?" The answer is that the RNase concentration is bundled as part of k_{mdec} (see the explanation for Equation 3.7).

The three assumptions described above—rate of zero, constant rate, and mass action kinetic rate—are by far the most common starting points for building an ODE-based model of a biological system. You will see that there are other ways to describe rates as well, but for now, simplifying the equation for protein concentration using these three assumptions is pretty straightforward. We use the mass action kinetics assumption for all three terms on the right-hand side of the equation:

$$\frac{d[Protein]}{dt} = k_{trl}[mRNA] - k_{pdec}[Protein] - k_{pdil}[Protein] \qquad (3.9)$$

Protein decay and dilution are simply proportional to the protein concentration, and because translation creates protein from mRNA, we assume that the translation rate k_{trl} is proportional to the mRNA concentration. Just as we assumed that the RNase concentration did not change much over time to obtain Equation 3.8 from Equation 3.7, we are assuming that the concentrations of ribosomes and protease are relatively stable and can therefore be bundled into our kinetic constants. Since both the decay and the dilution terms depend on the protein concentration, we can also combine these terms into one general "loss" term:

$$\frac{d[Protein]}{dt} = k_{trl}[mRNA] - k_{ploss}[Protein] \tag{3.10}$$

Our mRNA equation is somewhat harder to write as a result of our system's autorepression. We need to come up with an expression for the rate of mRNA production that reflects the idea that less mRNA will be produced if there is a lot of protein available. We do not know exactly what this function looks like yet, but we want to assert that protein concentration is the key player. For now, let's just say that the rate of transcription is a function of the protein concentration, $fxn_{trs}([Protein])$. We can then rewrite Equation 3.5 as:

$$\frac{d[mRNA]}{dt} = fxn_{trs}([Protein]) - k_{mloss}[mRNA] \tag{3.11}$$

Except for that as-yet-unknown function, Equations 3.10 and 3.11 are pretty well specified.

THE STEADY-STATE ASSUMPTION

Can we reduce this set of two equations to only one equation that retains the key information? One way to approach this simplification is to consider timescales, just as we did with our Boolean modeling. In Chapter 2, you learned that the production of mRNA and protein occurs at similar rates, but that mRNA decays in general more quickly than protein. This difference in the rate of loss means that if there is a perturbation to our system, the mRNA concentration will probably recover to a stable level more quickly than the protein concentration. As a result, we can consider what happens to the protein concentration when mRNA is considered to be at its **steady-state** concentration, which we will call $[mRNA]_{ss}$. Thus:

$$\frac{d[mRNA]_{ss}}{dt} = 0 = fxn_{trs}([Protein]) - k_{mloss}[mRNA]_{ss} \tag{3.12}$$

rearranges to:

$$[mRNA]_{ss} = fxn_{trs}([Protein])/k_{mloss} \qquad (3.13)$$

We can then reduce our two-equation system into a single equation, substituting this steady-state value into our protein equation:

$$\frac{d[Protein]}{dt} = k_{trl}[mRNA]_{ss} - k_{ploss}[Protein] \qquad (3.14)$$

$$\frac{d[Protein]}{dt} = \frac{k_{trl}}{k_{mloss}} fxn_{trs}([Protein]) - k_{ploss}[Protein] \qquad (3.15)$$

To simplify Equation 3.15, let us define another function, $fxn_{trl}([Protein])$:

$$fxn_{trl}([Protein]) = \frac{k_{trl}}{k_{mloss}} fxn_{trs}([Protein]) \qquad (3.16)$$

Notice that fxn_{trl} is also only a function of the protein concentration. Substituting our new function into Equation 3.15, we obtain:

$$\frac{d[Protein]}{dt} = fxn_{trl}([Protein]) - k_{ploss}[Protein] \qquad (3.17)$$

and now we have an equation that is written only in terms of a single variable: [Protein].

SOLVING THE SYSTEM WITHOUT FEEDBACK: REMOVAL OF ACTIVATOR

At this point, let's revisit the actual nature of our translation function fxn_{trl} (Equation 3.16). First, consider the case without autoregulatory feedback (Figure 3.1a). Remember, when activator is added to the system, the gene is expressed at a constant high rate; when the activator is removed, gene expression ceases. We can represent this scenario mathematically as:

$$fxn_{trl}([Protein]) = \begin{cases} 0 \ (\text{no activator}) \\ k_{trl,max} \ (\text{activator}) \end{cases} \qquad (3.18)$$

where $k_{trl,max} = (k_{trl} \cdot k_{trs,max})/k_{mloss}$ and $k_{trs,max}$ is the absolute maximum amount of transcription that can occur from the gene. There are two cases of primary interest: We either remove activator from an active system or we add activator to an inactive system. Let's consider each case in turn.

First we have the case in which the system has been expressing protein for a while and has reached a steady state; then, the activator is removed. Assume for now that removal of the inducer leads to immediate cessation of protein production. Thus, $fxn_{trl} = 0$, and:

$$\frac{d[Protein]}{dt} = -k_{ploss}[Protein] \tag{3.19}$$

This equation can be readily solved analytically (Sidebar 3.1):

$$[Protein](t) = [Protein]_{t=0}\, e^{-k_{ploss}t} \tag{3.20}$$

SIDEBAR 3.1 ANALYTICALLY SOLVING A SIMPLE ODE

Given an ODE describing a variable u such that:

$$\frac{du}{dt} = au$$

First, rewrite the ODE so that all of the terms involving u are on one side and terms involving t are on the other:

$$\frac{du}{u} = a\, dt$$

Next, perform the integration of both sides. An integral table may be helpful, and these are easy to find online:

$$\ln\left(\frac{u}{u_0}\right) = a(t - t_0)$$

Assuming that $t_0 = 0$, solving for u leads to:

$$u(t) = u_0 e^{at}$$

You can double-check your answer by differentiating:

$$\frac{du}{dt} = \frac{d(u_0 e^{at})}{dt} = a(u_0 e^{at}) = au$$

Notice that the solution involves an **exponential** component, in which e is raised to the power of at. Exponential terms will play a large role in our discussion of ODE models. The process of verifying that Equation 3.20 is the solution of Equation 3.19 is analogous.

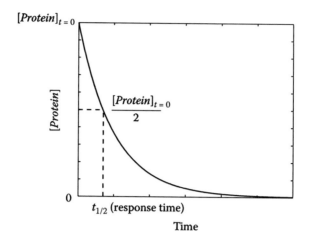

FIGURE 3.5 Removing the activator immediately inhibits transcription of our circuit, causing the protein concentration to decay over time. By assuming a solution with an exponential form, we can monitor the protein concentration at any given point in an experiment. The response time of the system occurs when the initial protein concentration has been halved.

From Equation 3.20, it is relatively easy to graph the protein concentration over time. As t increases, the value of the exponential term becomes smaller and smaller, approaching zero (Figure 3.5). This scenario is called an exponential decay, and you will see many of these kinds of terms as we continue.

KEY PROPERTIES OF THE SYSTEM DYNAMICS

I used MATLAB to construct Figure 3.5, but you could just as easily sketch it with three quick calculations and your intuition. These three calculations represent important aspects of the system dynamics: (1) where the system starts, (2) what happens at long times, and (3) how the system transitions from the initial to the final conditions.

First, you already know that the initial protein concentration is $[Protein]_{t=0}$, so you can plot a point at $(0, [Protein]_{t=0})$. Second, when the system reaches a steady state, $d[Protein]/dt = 0$, so in this case $[Protein]_{ss}$ must equal zero as well. Therefore, you can plot that at long times $[Protein] = [Protein]_{ss} = 0$.

Finally, you can use the equation to determine the time $t_{1/2}$ at which the initial protein concentration is decreased by half, so that

$[Protein]_{(t=t_{1/2})} = 1/2[Protein]_{(t=0)}$. This time is sometimes called the **response time** and is determined by substitution:

$$[Protein](t=t_{1/2}) = [Protein]_{t=0}\, e^{-k_{ploss}t_{1/2}} = \frac{[Protein]_{t=0}}{2} \qquad (3.21)$$

Simplifying, we find that:

$$t_{1/2} = \frac{\ln(2)}{k_{ploss}} \qquad (3.22)$$

Using these three points, and knowing that there is an exponential decay in Equation 3.20, you could draw Figure 3.5 by hand. Of course, you could also use MATLAB to draw the exact plot:

```
% set constants equal to one for illustrative purposes
Protein_0 = 1;
k_ploss = 1;
time = 0:0.1:6;
Protein_t = Protein_0 * exp(-k_ploss * time);
plot(time, Protein_t);
```

SOLVING THE SYSTEM WITHOUT FEEDBACK: ADDITION OF ACTIVATOR

Now, let's consider the case in which there is no gene expression and the activator is added such that gene translation suddenly occurs at its maximal rate $k_{trl,\,max}$. In this case, $[Protein]_{t=0} = 0$, and the steady-state solution is:

$$\frac{d[Protein]}{dt} = 0 = k_{trl,\,max} - k_{ploss}[Protein]_{ss} \qquad (3.23)$$

$$[Protein]_{ss} = \frac{k_{trl,\,max}}{k_{ploss}} \qquad (3.24)$$

The full analytical solution is not derived here, but it is common enough to find in an integral table; the solution is:

$$[Protein](t) = [Protein]_{ss}\left(1 - e^{-k_{ploss}t}\right) \qquad (3.25)$$

Once again, you see an exponential component as the protein concentration rises to its steady-state value (Figure 3.6). Notice also that the steady-state

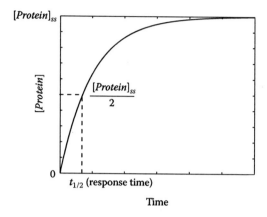

FIGURE 3.6 Dynamics of our system when addition of activator immediately drives translation to its maximal rate. Notice that the response time of this system is the same as the response time of the system in Figure 3.5.

solution of the system when activator is removed is equal to the initial condition of the system when activator is suddenly added and vice versa.

With Equation 3.25, you can also determine the response time of the activation of gene expression:

$$[Protein]_{(t=t_{1/2})} = [Protein]_{ss}\left(1 - e^{-k_{ploss}t_{1/2}}\right) = \frac{[Protein]_{ss}}{2} \quad (3.26)$$

Interestingly, the response time is the same as calculated in Equation 3.22:

$$t_{1/2} = \frac{\ln(2)}{k_{ploss}} \quad (3.27)$$

What does this mean? Recall that k_{ploss} is a combination of two terms, a dilution term and a decay term, and the dilution term is dominant. The dilution term is related to the observation that every cell division reduces the protein concentration in each daughter cell by half. In other words, the time it takes to divide the protein concentration in half is the time it takes for one cell to divide. Therefore, $t_{1/2}$ is simply the doubling time of the bacterium, and because doubling time is related to growth rate by:

$$doubling\ time = \frac{\ln(2)}{growth\ rate} \quad (3.28)$$

k_{ploss} is roughly equal to the growth rate.

COMPARISON OF MODELING
TO EXPERIMENTAL MEASUREMENTS

Now, let's look at some experimental measurements. Figure 3.7 shows the results of an experiment using the genetic construct in Figure 3.1a, where the system is suddenly induced from a state of no gene expression. You can see that the theory and experiment agree nicely.

Let's review what we have learned so far in this chapter. We focused on a construct without any feedback (Figure 3.1a) for which the input functions and rates are relatively simple. We discussed some critical assumptions, such as mass action kinetics, and relating timescales to each other. We examined the response time as well as the steady states. We also worked on building intuition: If you know a steady state and you know how quickly something responds, you can sketch a simple model, equations, and often even a plot on the back of an envelope.

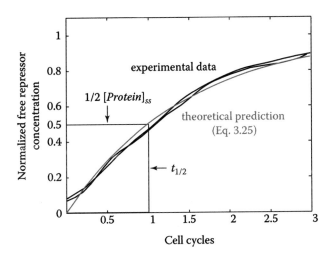

FIGURE 3.7 Experimental measurement of protein (repressor) concentration over time following sudden induction of a system without feedback (Figure 3.1a). *E. coli* cells were grown in batch culture and periodically monitored for GFP fluorescence, which served as a proxy for protein concentration. The experimental measurements (black lines) are in reasonable agreement with the concentrations predicted (red line) by the set of ODEs. Note that the normalized protein concentration is halved after one cell cycle, as predicted by Equations 3.27 and 3.28. (Modified from Rosenfeld, N., Elowitz, M. B., and Alon, U. *Journal of Molecular Biology* 2002, **323**(5): 785–793, with permission from Elsevier.)

ADDITION OF AUTOREGULATORY FEEDBACK

Let's see what happens when we add autoregulatory feedback, using the system in Figure 3.1b as our example. We apply some of the same steps learned previously to determine why autoregulatory feedback loops are so common in *E. coli*'s transcriptional regulatory network.

The key challenge in this case will be to determine fxn_{trs} and its dependence on the protein concentration. To begin thinking about this, look at the depiction in Figure 3.8 of the interaction between the protein and the operator region. Free protein binds to free DNA to form a complex, which can also dissociate to release protein and free DNA. The association and dissociation processes both have kinetic constants associated with them.

PRACTICE PROBLEM 3.2

Given the system in Figure 3.8, write an ODE that represents the concentration of bound DNA over time. State any assumptions that you make. How is the bound DNA concentration related to the free DNA concentration under steady-state conditions?

SOLUTION

Bound DNA is produced by association of free protein and DNA and lost by dissociation of the protein-DNA complex. Your initial equation will therefore look something like:

$$\frac{d[DNA_{bound}]}{dt} = Rate_{association} - Rate_{dissociation} \tag{3.29}$$

To further specify the rates, assume that association and dissociation follow mass action kinetics:

$$\frac{d[DNA_{bound}]}{dt} = k_a [Protein][DNA_{free}] - k_d [DNA_{bound}] \tag{3.30}$$

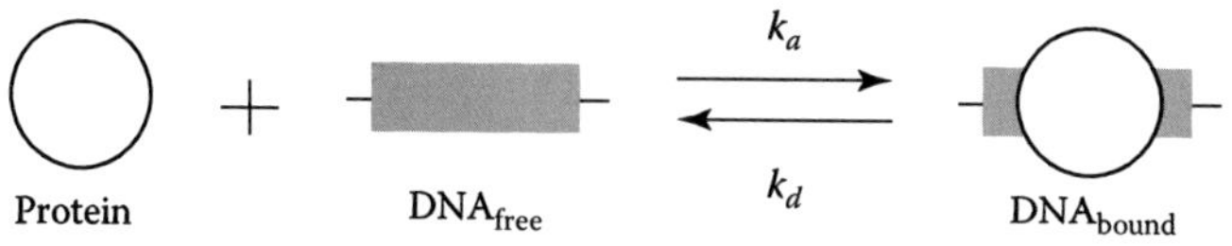

FIGURE 3.8 The dynamics of binding of a transcription factor (protein) to free DNA include association (k_a) and dissociation (k_d) kinetics. The transcription factor binding site is represented as a gray box on the DNA (straight line).

At steady state:

$$\frac{d[DNA_{bound}]_{ss}}{dt} = 0 = k_a[Protein][DNA_{free}]_{ss} - k_d[DNA_{bound}]_{ss} \qquad (3.31)$$

Rearranging, we find that:

$$[DNA_{bound}]_{ss} = \frac{k_a}{k_d}[Protein][DNA_{free}]_{ss} \qquad (3.32)$$

If we define an **equilibrium** dissociation constant $K = k_d/k_a$, we can simplify the expression to:

$$[DNA_{bound}]_{ss} = \frac{[Protein]}{K}[DNA_{free}]_{ss} \qquad (3.33)$$

Now, we have a relationship between the concentrations of free and bound DNA that depends on the protein concentration and the binding affinity between the protein and DNA. A strong affinity means that DNA and protein are likely to associate and not let go of each other; K is therefore small because $k_a \gg k_d$. Similarly, a weak affinity leads to a high K.

We are after something a little different, however: We would like to know how much of the total DNA is active and can be transcribed into mRNA. Our transcription factor (TetR) is a repressor, so the free DNA is the active DNA. What is the ratio of free DNA to total DNA? Here, another conservation equation can be useful. We assume that the total amount of DNA is constant and that the DNA can only exist as free or bound. Then, we can write:

$$[DNA_{total}] = [DNA_{bound}] + [DNA_{free}] \qquad (3.34)$$

Equation 3.34 is valid at any point in time, including at steady state. We can therefore substitute Equation 3.33 into Equation 3.34 to obtain

$$[DNA_{total}] = \frac{[Protein]}{K}[DNA_{free}]_{ss} + [DNA_{free}]_{ss} \qquad (3.35)$$

Simplifying yields:

$$\frac{[DNA_{free}]_{ss}}{[DNA_{total}]} = \frac{1}{1 + \dfrac{[Protein]}{K}} \qquad (3.36)$$

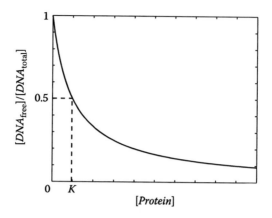

FIGURE 3.9   The fraction of free DNA is determined by the amount of protein available to bind the DNA. Note the definition of K, the equilibrium dissociation constant, when half the DNA in the system is bound by protein.

Take a minute to understand what this equation means, using Figure 3.9 as a guide. If there is no protein available to bind DNA ($[Protein] = 0$), then the concentration of free DNA is equal to the total amount of DNA because all DNA is free. If the protein concentration is very high such that $[Protein] \gg K$, then the ratio of free DNA to total DNA is approximately zero, and virtually all DNA is bound. If the protein concentration is equal to K, then half of the DNA is free and the other half is bound. In other words, the fraction of active DNA is completely, and in this case negatively, dependent on the protein concentration for this system.

With the fraction of active DNA determined as a function of the protein concentration, we can identify a strategy for defining fxn_{trl}. Specifically, we can assume that the ratio of protein production to the maximum production rate is equal to the fraction of active DNA to the total amount of DNA:

$$fxn_{trl} = k_{trl,max} \frac{\left[DNA_{free}\right]_{ss}}{\left[DNA_{total}\right]} \tag{3.37}$$

Therefore:

$$fxn_{trl} = k_{trl,max} \frac{1}{1 + \dfrac{[Protein]}{K}} \tag{3.38}$$

and our ODE now looks like this:

$$\frac{d[Protein]}{dt} = \frac{k_{trl,max}}{1+[Protein]/K} - k_{ploss}[Protein]. \qquad (3.39)$$

Notice that the mass action assumption-based approach to modeling protein-DNA binding led to a more complicated term for protein production in Equation 3.39.

COMPARISON OF THE REGULATED AND UNREGULATED SYSTEMS

Now that the ODE is defined, let's compare this system (Figure 3.1b) to the one without feedback (Figure 3.1a). To do this, we will find the steady-state solution, the response curve for a sudden activation of gene expression (the **step response**), and the response time $t_{1/2}$.

The steady-state solution is determined from Equation 3.39 setting $d[Protein]/dt = 0$ and then solving for $[Protein]_{ss}$. In this case, we end up with a quadratic equation:

$$\frac{k_{ploss}}{K}[Protein]_{ss}^2 + k_{ploss}[Protein]_{ss} - k_{trl,max} = 0 \qquad (3.40)$$

which we simplify by multiplying both sides of the equation by K/k_{ploss} and then obtain the solution:

$$[Protein]_{ss} = \frac{-K+\sqrt{K^2+4\dfrac{K\,k_{trl,max}}{k_{ploss}}}}{2} \qquad (3.41)$$

We can simplify Equation 3.41 with one more assumption: very strong autorepression. As mentioned, this scenario means that $k_a \gg k_d$; K is very small and can be neglected with respect to other terms. Applying this assumption, both inside and outside the radical, and dividing both the numerator and denominator by 2 yields:

$$[Protein]_{ss} = \sqrt{\frac{K\,k_{trl,max}}{k_{ploss}}} \qquad (3.42)$$

Recall from Equation 3.24 that the system without feedback had a steady-state protein concentration equal to $k_{trl,max}/k_{ploss}$. Given that the value of K is expected to be very small, you would expect that $[Protein]_{ss}$ is significantly smaller with feedback than without. Indeed, Rosenfeld and colleagues (who carried out the experimentation with these circuits) calculated the steady-state protein concentration using these equations and determined approximate parameter values for bacterial repressors from the literature, estimating a steady-state protein concentration of ~4,000 proteins per cell in the case of no autoregulation and ~200 proteins per cell with autoregulation—a 20-fold difference!

To find the response time, we have to integrate Equation 3.39. You can refer to an integral table if you would like proof, but the solution is:

$$t([Protein]) - t([Protein]_0)$$

$$= -\frac{1}{2k_{ploss}} \left[\ln\left(\left([Protein] - [Protein]_{ss}\right)\left([Protein] + K + [Protein]_{ss}\right)\right) \right.$$

$$\left. + \frac{K}{K + 2[Protein]_{ss}} \ln\left(\frac{[Protein] - [Protein]_{ss}}{[Protein] + K + [Protein]_{ss}} \right) \right] \Bigg|_{[Protein]_0}^{[Protein]} \tag{3.43}$$

We can make a few assumptions to simplify this expression. For example, let's say that the starting time and protein concentration $[Protein]_0$ are both equal to 0. As before, we also assume strong autorepression and therefore that K is very small compared to the other terms. These assumptions enable us to reduce Equation 3.43 to:

$$t([Protein]) = -\frac{1}{2k_{ploss}} \ln\left(\frac{[Protein]_{ss}^2 - [Protein]^2}{[Protein]_{ss}^2} \right) \tag{3.44}$$

Solving for the ratio of protein concentration to steady-state protein concentration:

$$\frac{[Protein]}{[Protein]_{ss}} = \sqrt{1 - e^{-2k_{ploss}t}} \tag{3.45}$$

PRACTICE PROBLEM 3.3

Determine the response time $t_{1/2}$ for the autoregulated system in Figure 3.1b.

SOLUTION

We determine the response time with Equation 3.45 by setting the protein concentration equal to one-half of its steady-state value. Substitution yields:

$$\frac{1}{2} = \sqrt{1 - e^{-2k_{ploss}t_{1/2}}} \tag{3.46}$$

and by rearranging and solving, we obtain:

$$t_{1/2} = \frac{\ln(4/3)}{2k_{ploss}} \tag{3.47}$$

Now that we have investigated the unregulated and autoregulated systems, we can compare some of the key features. Figure 3.10 compares protein concentration with respect to the steady-state protein concentration over time (the perturbation response or step response), the steady-state protein concentration, and the response time for the unregulated and strongly autoregulated models. The results in each case appear similar, but the perturbation response leads to a somewhat shorter response time— about a five-fold difference if you work out the math. The steady-state protein level is also lower, as discussed previously.

	Without negative autoregulation	With negative autoregulation
Step response	$1 - e^{-k_{ploss}t}$	$\sqrt{1 - e^{-2k_{ploss}t}}$
Steady-state protein level	$\dfrac{k_{trl,\,max}}{k_{ploss}}$	$\sqrt{K\dfrac{k_{trl,\,max}}{k_{ploss}}}$
Response time	$\dfrac{\ln(2)}{k_{ploss}}$	$\dfrac{\ln(4/3)}{2k_{ploss}}$

FIGURE 3.10 Comparison of the key features of gene circuits without and with strong negative autoregulation. The circuits from Figure 3.1 have similar features, but the feedback system has a lower steady-state protein concentration as well as a shorter response time to perturbation.

The incorporation of autoregulation thus leads to a lower steady-state protein level and a faster response to perturbation such that the steady state is reached more quickly. In other words, feedback makes the circuit more responsive and efficient. In contrast, the circuit without feedback will either be producing more protein than is necessary or will be taking a relatively long time to respond to perturbations.

Are these predicted differences actually borne out in the laboratory? The answer turns out to be yes. We have already discussed how the steady-state protein levels are significantly lower in autoregulated circuits than in unregulated circuits. The same team of experimentalists measured the response times of both circuits in Figure 3.1; their results appear in Figure 3.11, which is similar to Figure 3.7. First, note that the *y* axis is scaled for comparative purposes—we expect the two circuits to have different steady-state values, but they have been normalized. The experimental data are shown near the theoretical predictions, and indeed, the response time is significantly shorter in the autoregulated circuit.

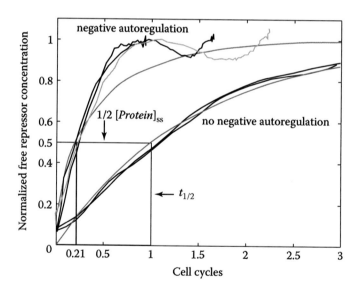

FIGURE 3.11 Experimental measurement of protein (repressor) concentration over time following sudden induction of systems without or with strong negative autoregulation. ODEs were used to predict the normalized protein concentrations for the system without feedback and the system with feedback (red lines). Experimental measurements are depicted with black and gray lines. (Modified from Rosenfeld, N., Elowitz, M. B., and Alon, U. *Journal of Molecular Biology* 2002, **323**(5): 785–793, with permission from Elsevier.)

Now that we have looked at two very different methods of approaching the same problem—autoregulatory negative feedback in transcriptional regulation—I think you will agree that the methods are quite different! I hope you have seen that these tools contribute several elements that we could not have deduced with our Boolean modeling. For example, we could not have calculated and compared response times and steady states. ODEs are a powerful tool in your systems biology tool kit!

By covering the analytical solution of ODEs in this chapter, I have also been trying to help you build more biological and computational intuition. Biologically, we unraveled another layer of the feedback scenario to gain further insight about the role that feedback plays in transcription and, as you will see further in this book, in other systems as well.

Computationally, I hope you saw how powerfully certain assumptions enabled us to create and solve these equations. The key assumptions were conservation of mass, mass action kinetics, strong affinity binding, and steady-state conditions, and they are going to show up again and again throughout this text. The next chapter focuses even more on intuition and back-of-the-envelope calculation using graphical approaches to solve ODEs.

CHAPTER SUMMARY

The most common approach for describing models of biological systems is to write a set of ODEs. A convenient way to conceptualize ODEs is to draw a compartment model diagram, which shows the variables as blocks; the processes that change the variable values are given as arrows pointing toward or away from the blocks. Once the variables and processes have been identified, the processes must be specified in more detail. Three common assumptions that can aid in specification are:

1. The rate at which the process occurs is very small compared to other processes and therefore can be neglected.

2. The rate at which the process occurs is constant and does not change based on any other variable or rate.

3. The rate at which the process occurs follows mass action kinetics, meaning that the rate is proportional to the product of the concentrations of the reactants.

Only relatively small and uncomplicated sets of ODEs can be solved analytically, but the analytical solution can give you important insights into how a system works. For this reason, it is often worth the effort to simplify your model, as we demonstrated by reducing our two-equation model to a single equation, in order to build intuition. This simplification depended on a further common assumption, based on the consideration of timescales: We assumed that one of our variables reached a steady state faster than the other variable. This assumption enabled us to solve for that steady-state value and substitute into the other equation.

In keeping with our focus on negative autoregulation in gene expression, we considered a pair of complementary synthetic gene circuits with one difference: One was negatively autoregulated, and the other was not. The analytical approach enabled us to derive the properties of the two circuits, including the response time, the step response, and the steady-state expression level. We found that the step responses were related, but different; in the autoregulated circuit, the response time was significantly shorter and the steady-state protein level was lower than in the nonautoregulated circuit. This observation suggests that autoregulation enables cells to respond quickly to changes in their environment by tuning the expression of fewer proteins. Finally, we showed that our theoretical analysis matched well with reported experimental measurements.

In summary, analytical solutions of ODEs add another layer of dynamical information and complexity to our modeling tool kit. We cannot see all of the possible solutions at once, as we did with Boolean models. Nevertheless, the gain in our ability to make quantitative predictions by using ODEs often outweighs this drawback. Moreover, techniques exist to simultaneously visualize multiple ODE systems, as you will see in the next chapter.

RECOMMENDED READING

Alon, U. *An Introduction to Systems Biology: Design Principles of Biological Circuits.* Boca Raton, FL: Chapman & Hall/CRC Press, 2007.

Bolouri, H. *Computational Modeling of Gene Regulatory Networks—A Primer.* London: Imperial College Press, 2008.

Rosenfeld, N., Elowitz, M. B., and Alon, U. Negative autoregulation speeds the response times of transcription networks. *Journal of Molecular Biology* 2002, **323**(5): 785–793.

Strang, G. *Linear Algebra and Its Applications.* 4th edition. Stamford, CT: Brooks Cole, 2005.

PROBLEMS

PROBLEM 3.1
Cell Growth

Cells growing in a nutrient-rich environment (such as *E. coli* growing in a flask full of rich broth) are observed to grow exponentially; their growth can be represented as:

$$\frac{dX}{dt} = \mu X$$

where X is the cellular biomass concentration, and μ is the growth rate.

a. Write the analytical solution of the growth equation, assuming an initial cellular biomass concentration of X_0.

b. Determine the relationship between the growth rate and the doubling time: the time at which $X(t) = 2X_0$. Relate this relationship to the concept of response time that you read about in this chapter.

c. Now, imagine that you are growing cells in a flask and measuring the concentration of cells over time. From your measurements, you determine a growth rate that you call μ_{flask}. However, when you watch these cells growing under a microscope, you notice that some of the cells lyse and die at a rate proportional to the biomass concentration, with rate constant k_{death}. How would you change the ODE in your cell growth model to account for the death of cells over time?

d. Write the analytical solution of the ODE in (c).

e. What condition is required for this system to reach steady state?

f. Going back to your microscope, you decide to measure the growth rate of individual cells by recording the time between cell division events in a single cell. Based on your model from (c), how would you expect your new single-cell measurements to compare to the growth rate that you measured in (a)?

PROBLEM 3.2
Negative Autoregulation

In this problem, we focus on the same circuits discussed in this chapter, but with more detail.

First, we consider the system without autoregulation.

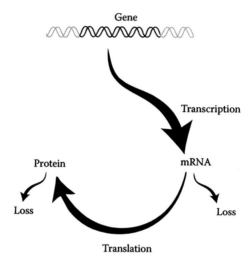

a. Derive a single ODE for the change in protein concentration over time, given rate constants for the rates of mRNA and protein loss (k_{mloss} and k_{ploss}, respectively), translation (k_{trl}), and maximum transcription ($k_{trs,\,max}$). List the assumptions that were required to derive this equation.

b. Using the equation from (a), calculate k_{ploss} and the steady-state protein concentration in both micromoles and the number of proteins per cell. You may assume that $[Protein]_{(t=0)} = 0\ \mu M$, and that transcription is suddenly induced by the addition of an activator. You will also need the following constants:

Division time of a cell = 35 min

$k_{trl,\,max}$ (maximum protein production rate) = 0.0015 μM/s

1 μM protein concentration ≈ 1,000 protein molecules/cell

c. Write the analytical solution of the equation you derived in (a) and generate a graph of the change in protein concentration over time based on this equation and the constants given.

d. Now, use the MATLAB ODE solver ode45 to plot the numerical solution to the ODE (type help ode45 in the command line in MATLAB). How does this plot compare to your solution to (c)?

We consider now the case of negative autoregulation.

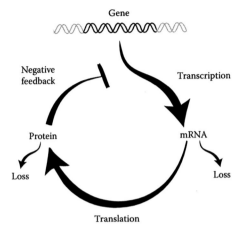

Here, the protein can bind to the operator, and transcription only occurs in the absence of protein binding. As before, K is the ratio of the rate of protein dissociation from the operator to the rate of protein association with the operator, and the rate of protein production is

$$\frac{k_{trl,max}}{\left(1+\dfrac{[Protein]}{K}\right)}$$

e. Calculate the steady-state protein concentration (in micromoles and proteins/cell) and the response time of the system. Be careful with units. The value for K is 15 nM.

f. Plot the changes in protein concentration over time using ode45. Again, the initial protein concentration is zero. Put this plot on the same axes as your output for (c).

g. Compare the values and plots from (e) and (f) to those for the case without regulation in (b) and (c). Comment briefly on the possible advantages of this difference in behavior in biological terms.

PROBLEM 3.3
Positive Autoregulation

In Problem 2.1, we considered an autoregulatory loop with positive, instead of negative, feedback. We now consider this circuit in more detail using ODEs.

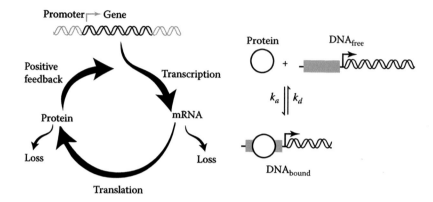

On the right side of the figure for this problem is a diagram of the binding between the protein and free DNA (DNA_{free}) to form a complex, DNA_{bound} (you have already seen this diagram in Figure 3.8). In this case, only the bound DNA can be transcribed.

a. Write an ODE for the change in DNA_{free} over time in terms of DNA_{free}, DNA_{bound}, and *Protein*, together with the association and dissociation rate constants k_a and k_d.

b. What assumption did you make to write the equation in (a)? Give an example of a case for which this assumption does not hold.

c. Use your answer from (a) to derive an expression for the ratio of $[DNA_{bound}]$ to $[DNA_{total}]$. Your answer should be in terms of *Protein* and K, where $K = k_d/k_a$.

d. Use your answer to (c) together with Equation 3.17 to write a single ODE that describes the change in protein concentration over time.

e. Write an equation for the steady-state protein concentration in terms of K, k_{ploss}, and $k_{trs, max}$. How does this new equation compare to the equations for the system with no regulation?

f. Using MATLAB, plot a graph of the dynamics for (1) the equation you used in (d), with $k_{trl, max} = 0.0008$ µM/s, $k_{ploss} = 0.0001925$/s, and $K = 0.010$ µM; (2) the same as in (1), but with $K = 1$ µM; and (3) the system without positive regulation. Plot them all on the same graph for 0–5 h. How do the three compare? Note that for (3) you can use the initial conditions $t = 0$ h and $[Protein] = 0$ µM, but for (1) and (2), you will prefer initial conditions such as $t = 0$ h and $[Protein] = 0.1$ µM. Why?

PROBLEM 3.4
Negative Autoregulation with Cooperativity

A common variation on our negative autoregulatory circuit is that, often, more than one binding site for the same protein occurs in the promoter region of the DNA. Therefore, more than one molecule of the same species can bind to the DNA and more strongly repress expression. This phenomenon is known as **cooperativity**.

We now analyze a negative autoregulatory circuit with cooperativity. In this case, two proteins can bind to the DNA, and both must be bound to the DNA to repress transcription.

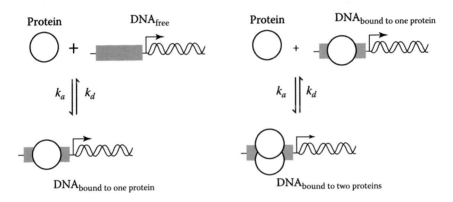

The reactions describing the interactions between DNA and protein are:

$$protein + DNA_{free} \underset{k_d}{\overset{k_a}{\rightleftharpoons}} DNA_{bound\ to\ one\ protein}$$

$$protein + DNA_{bound\ to\ one\ protein} \underset{k_d}{\overset{k_a}{\rightleftharpoons}} DNA_{bound\ to\ two\ proteins}$$

a. From these reactions, derive an equation for $[DNA_{free}]/[DNA_{total}]$ in terms of protein concentration and K.

b. The equation that you will see most often in the literature as the answer to (a) looks like this:

$$\frac{\left[DNA_{free}\right]}{\left[DNA_{total}\right]} = \frac{1}{1 + \left(\dfrac{[Protein]}{K}\right)^{H}}$$

where H is the Hill coefficient, which expresses the amount of cooperativity in the system. To obtain a solution in this form, you will need to cancel one term from your solution. Where does this term come from? Explain why it is reasonable to cancel this term in biological terms.

c. What is the value of H for our system? What happens when H is equal to 1? To what do you think higher values of H correspond?

d. For $H = \{1, 2, 3, 4, 5\}$, plot the ratio $[DNA_{free}]/[DNA_{total}]$ versus $[Protein]$, where $[Protein]$ varies from 0 to 0.05 μM. Let $K = 10$ nM. How do the curves change as H increases? What does this change mean in biological terms? What implications could this change have for modeling a cooperative reaction?

e. Substitute the equation given in (b) into Equation 3.17 and plot the changes in $[Protein]$ over time for this case of negative autoregulation and cooperativity. Initially, the system contains no protein, and there is a sudden increase in $k_{trl, max}$ to 0.0008 μM. The value for k_{ploss} is 0.0001925/s. Use ode45. Show lines for $H = \{1, 2, 3, 4, 5\}$. How does the steady-state protein concentration change as H increases? Explain this trend in terms of the biology of the system.

CHAPTER 4

Variation: Graphical Analysis

LEARNING OBJECTIVES

- Generate plots of X versus dX/dt for single ODE systems

- Calculate fixed points and determine their stability using vector fields

- Draw time-course plots from arbitrary initial conditions using graphical analysis

- Understand how changing parameters can lead to bifurcation

When I ask my students about the most surprising moments they experienced in my systems biology course, the power of Boolean analysis is usually on their list. Many of them are particularly caught off guard by the number of conditions that you can examine simultaneously using this approach. By comparison, our ODE-based analytical approach is relatively limited; we only looked at a couple of initial conditions and had to make pretty specific assumptions to get there.

What makes graphical analysis so exciting is that, once again, we can examine many initial conditions at once to see how the system will behave. We can also deduce the ranges of initial conditions for which the response will be similar, and we can even characterize how parameter changes lead to dramatic, qualitatively different, altered responses.

Also, graphical analysis is *beautiful*. There is not a single figure I will show you here that I would not love to put on my wall at home! This aesthetic quality also helps your intuition; by visualizing many responses at once, your eyes and mind will help you to identify patterns that you can remember.

REVISITING THE PROTEIN SYNTHESIS ODES

Let's start with our ODE descriptions of our circuits in which the protein concentration varies without (Figure 3.1a) and with (Figure 3.1b) feedback, taken from Equations 3.23 and 3.39, respectively:

$$\frac{d[Protein]}{dt} = k_{trl,\,max} - k_{ploss}[Protein] \tag{4.1}$$

$$\frac{d[Protein]}{dt} = \frac{k_{trl,\,max}}{1+[Protein]/K} - k_{ploss}[Protein] \tag{4.2}$$

Remember that $k_{trl,\,max}$ is a constitutive (constant) production rate, and k_{ploss} is a rate constant that describes the relationship between protein concentration and protein loss over time, based on an assumption of mass action kinetics. The more complicated production term in Equation 4.2 also assumes mass action kinetics, but here it reflects the binding of repressor protein to DNA to form an inactive complex and the dissociation of that complex.

We solved these equations in Chapter 3. The solution to Equation 4.1 is:

$$[Protein](t) = [Protein]_{ss}\left(1 - e^{-k_{ploss}t}\right) \tag{4.3}$$

where

$$[Protein]_{ss} = \frac{k_{trl,\,max}}{k_{ploss}} \tag{4.4}$$

We then plotted Equation 4.3 (Figure 3.6) to visualize the dynamics of the system responding to a sudden increase in expression, starting at $[Protein] = 0$; this plot is shown again in Figure 4.1, but notice the new annotations in black and gray. First, the quantity $k_{trl,\,max} \cdot t$ is plotted in gray. At small values of t, when the response of the circuit has just started, the gray dashed line is a good approximation of the response.

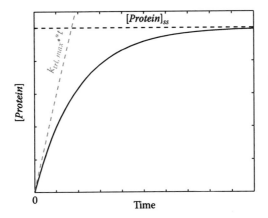

FIGURE 4.1 The response of our system (without feedback) to a sudden increase in protein expression (See also Figure 3.6). The initial increase in protein concentration closely approximates the translation rate (dashed light gray), but eventually, the system plateaus to a steady state (dashed black).

Reexamination of Equation 4.1 reveals that when not much time has passed, the protein concentration is low and the loss term is close to zero, leaving an equation in which $d[Protein]/dt \sim k_{trl, max}$. At larger values of t, the response is well approximated by the black dashed line, which is simply the steady-state protein concentration. At early times, the rate of change is therefore approximately $k_{trl, max}$, and at late times, the rate of change is zero.

PLOTTING X VERSUS DX/DT

Let's look at the rates of change in another way: still graphically, but this time plotting the variable value versus the variable's change over time. For our case, this strategy means plotting the protein concentration against the protein concentration over time, or $[Protein]$ versus $d[Protein]/dt$. The plot in Figure 4.2 is a straight line, with the y intercept at $k_{trl, max}$ and a slope of $-k_{ploss}$. You could also determine these values by examining Equation 4.1, but you can gain insight about the system by visualizing it as a graph and comparing it to Figure 4.1.

In particular, notice that $k_{trl, max}$ is the instantaneous rate of change for the protein concentration over time when the protein concentration is equal to zero—our initial conditions in Figure 4.1. You can also determine the steady-state protein concentration simply by finding the spot on the graph where $d[Protein]/dt = 0$.

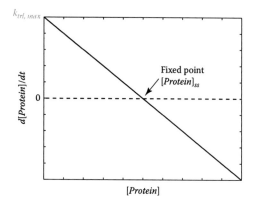

FIGURE 4.2 Comparing the change in protein concentration over time to the protein concentration reveals a linear relationship. As in Figure 4.1, the initial change in protein concentration over time mirrors the translation rate. The steady-state protein level occurs at $d[Protein]/dt = 0$; this point is called a "fixed" point.

FIXED POINTS AND VECTOR FIELDS

Points where the derivative of the variable is equal to zero are called **fixed points** of the system. One notable aspect of Figure 4.2 is that there is one and only one fixed point.

Fixed points can be either stable or unstable, with a critical impact on the system's behavior. Think of stability the way we discussed in previous chapters, beginning with Boolean analysis: If the system reaches a stable state, it will remain in that state. Likewise, a system in an unstable state moves away from that state.

In our case, you might already guess that the fixed point is stable because the trajectory in Figure 4.1 continues to approach it over time, but let's take this opportunity to show how you could infer this if you did not already know—by drawing a **vector field**. The vector field is an indicator of how much and in which direction the derivative is changing at a given variable value. We use our vector field to answer two questions. First, for a given protein concentration, will the rate of change be positive or negative? Second, how fast will the concentration change in that positive or negative direction?

The vector field for our system is shown in Figure 4.3 as a set of red arrows on the $d[Protein]/dt = 0$ line. We draw it by looking at the sign and magnitude of the $d[Protein]/dt$ value for a given protein concentration; for example, at $[Protein] = 0$, $d[Protein]/dt = k_{trl, max}$, a positive value that will result in a higher protein concentration over time. We therefore draw an arrow in the positive direction beginning at $[Protein] = 0$.

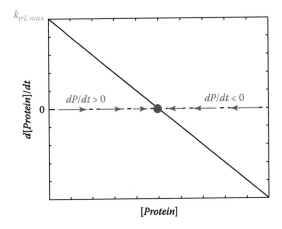

FIGURE 4.3   The vector field for the system in Figure 4.2. The arrows indicate the magnitude and direction of the change of $d[Protein]/dt$ as we approach the stable fixed point (filled circle). This vector field allows us to predict the system's response to any set of initial conditions.

If we move slightly to the right on the x axis to higher protein concentrations, we see that $d[Protein]/dt$ is still positive, but not as large as it was for $[Protein] = 0$. Accordingly, we draw another arrow, still pointing in the positive direction, but not as long (sometimes people make the arrow the same length as the value of $d[Protein]/dt$, but for now let's just look at the arrow lengths relative to each other). As we move slightly farther to the right, we find a relatively small but positive value of $d[Protein]/dt$ and draw a short, right-pointing arrow to reflect it. Finally, there is no change in protein concentration at the fixed point, so let's draw a circle around that point. The stability of fixed points is often denoted by marking stable fixed points with filled circles and unstable fixed points with empty circles, so we fill in our fixed point as well.

This next concept is important: The vector field that we've drawn is so far an equivalent way of representing the trajectory of the system represented in Figure 4.1. Beginning at $[Protein] = 0$, the rate of change is maximal, after which it steadily and linearly decreases until the protein concentration reaches a steady value of $d[Protein]/dt = 0$. This decrease in rate produces the concave-down shape of the trajectory in Figure 4.1.

Now, let's complete the vector field and examine the conditions under which the protein concentration is greater than the steady-state level, $[Protein] > [Protein]_{ss}$. Starting at the right side of the x axis, we see high values for the protein concentration, $d[Protein]/dt < 0$, so we draw arrows

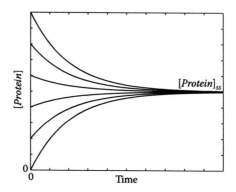

FIGURE 4.4   Sample trajectories of the system in Figure 4.2 given different initial starting conditions. All trajectories converge on the steady-state protein level, which appeared as a fixed point in Figure 4.3.

pointing to the left, and again the arrows become shorter as we approach the steady-state value until we reach the steady state.

FROM VECTOR FIELDS TO TIME-COURSE PLOTS

From this vector field, we can now predict the response of the system to any set of initial conditions. Specifically, if the system begins with [Protein] < [Protein]$_{ss}$, the protein concentration will increase at a steadily decreasing rate over time until it reaches the steady-state level, producing a concave-down trajectory. If the system starts at [Protein] > [Protein]$_{ss}$, the protein concentration will decrease at a steadily decreasing rate until it reaches that same steady-state level, producing a concave-up trajectory. Since the trajectories on either side approach the fixed point, we call it a stable state.

Some sample trajectories appear in Figure 4.4 above. I used MATLAB to draw this figure, but I want you to see that it is not necessary; likewise, you do not need the analytical solution. You could draw this figure by hand and amaze your friends with what you know about the vector field!

NONLINEARITY

In the first example, we looked at a case in which there was a linear relationship between [Protein] and d[Protein]/dt. Now, let's look at a slightly altered version of Equation 4.1, incorporating **nonlinearity** with a second-power term:

$$\frac{d[Protein]}{dt} = k_{trl,max} - k_{ploss}[Protein]^2 \qquad (4.5)$$

Such an equation may be appropriate if, for example, protein dimerization were required for the complex to be degraded (these systems do exist). You can see from Equation 4.5 that if $[Protein] = 0$, then $d[Protein]/dt = k_{trl, max}$, just as previously indicated. The steady-state solution, however, is slightly different:

$$[Protein]_{ss} = \pm\sqrt{\frac{k_{trl, max}}{k_{ploss}}} \qquad (4.6)$$

In reality, the protein concentration must be positive, but for learning purposes, let's consider all possible solutions of the equation. The plot of Equation 4.5 (Figure 4.5) crosses the y axis at $k_{trl, max}$ just as before, but now there are two fixed points, one at the positive solution of $[Protein]_{ss}$ and the other at the negative solution.

PRACTICE PROBLEM 4.1
Draw the vector field and determine the stability of each fixed point in Figure 4.5. Then, plot the protein concentration over time under the following initial conditions: (1) $[Protein]_0 = 0$; (2) $[Protein]_0 < -[Protein]_{ss}$; (3) $-[Protein]_{ss} < [Protein]_0 < 0$; (4) $0 < [Protein]_0 < +[Protein]_{ss}$; (5) $[Protein]_0 > +[Protein]_{ss}$.

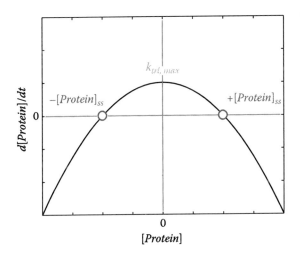

FIGURE 4.5 Plotting Equation 4.5 to visualize a nonlinear relationship between protein concentration and the change in protein concentration over time. Although two fixed points appear in the figure, protein concentration must be positive in living cells.

SOLUTION

The vector field appears in Figure 4.6. At the left side of the plot, $d[Protein]/dt$ is very negative, so the protein concentration continues to drop. The rate of change increases but remains negative as the leftmost fixed point is approached.

Just to the right of that fixed point, the rate of change is slow but positive and increases until it reaches a maximum value at $[Protein] = 0$. As $[Protein]$ becomes more positive, the rate decreases until it becomes zero at the rightmost fixed point. At values that are more positive, the rate again becomes negative, pointing toward that fixed point.

To determine the stability of each fixed point, notice that the vectors on either side of the rightmost fixed point are pointing toward that fixed point, just as we saw in Figure 4.3. This point is therefore a stable state of the system. In contrast, the vectors on either side of the leftmost fixed point are directed away from that point, resulting in an unstable state.

You can easily sketch the responses to any of the initial conditions using the vector field in Figure 4.6; some examples appear in Figure 4.7. Notice that time courses beginning near $-[Protein]_{ss}$ all move away from that fixed point; time courses beginning near $+[Protein]_{ss}$ move toward that value.

If we focus on the time courses that start at protein concentrations between 0 and $-[Protein]_{ss}$, we see that these trajectories begin by moving away from the unstable fixed point at an increasing rate as they

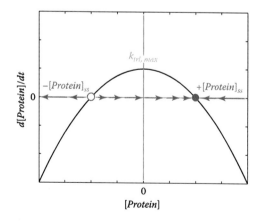

FIGURE 4.6 The vector field for the system in Practice Problem 4.1. The left fixed point is unstable, as indicated by the arrows facing away from the point.

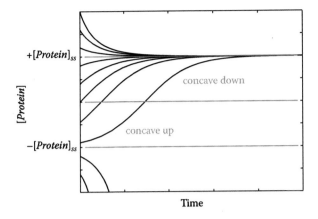

FIGURE 4.7 Responses of the system in Practice Problem 4.1, given different starting conditions. The inflection point in Figure 4.6 is indicated at [Protein] = 0, and the positive and negative fixed points are indicated at [Protein] = +/−[Protein]$_{ss}$.

approach [Protein] = 0. The shape of this curve, a slower change in [Protein] that increases over time, can be described as "concave up." After the time courses cross the [Protein] = 0 line, however, they begin moving toward the stable fixed point with a decreasing rate, resulting in a concave-down shape. The point at which the dynamics switch from concave up to concave down is called an **inflection point.**

BIFURCATION ANALYSIS

Changing the values of key parameters in an ODE can have a dramatic effect on the dynamics of your system, and graphical methods can powerfully illustrate these effects. Practice Problem 4.2 gives a pertinent example.

PRACTICE PROBLEM 4.2

Repeat the steps performed in Practice Problem 4.1 for $k_{trl, max} < 0$ and $k_{trl, max} = 0$. Draw vector fields, identify and characterize the fixed points, and sketch time courses for representative initial conditions. Don't use MATLAB until you've tried it by hand! Do either of these values of $k_{trl, max}$ have any biological relevance?

SOLUTION

When $k_{trl, max}$ is negative (Figure 4.8, top), there are no fixed points, but [Protein] = 0 remains an inflection point where d[Protein]/dt switches from decreasing in magnitude to increasing in magnitude.

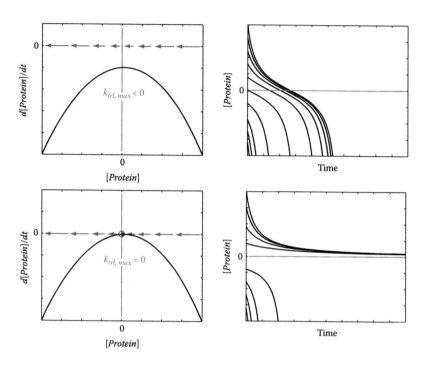

FIGURE 4.8 The value of $k_{trl,max}$ influences the number of fixed points (Practice Problem 4.2). Top, when $k_{trl,max}$ is negative, there are no fixed points. Bottom, when $k_{trl,max} = 0$, a half-stable point occurs at $[Protein] = 0$.

When $k_{trl,max} = 0$ (corresponding to complete repression of gene expression; Figure 4.8, bottom), there is only one fixed point; this point is unusual because the vector field on the left side of the fixed point is directed away, while the vector field on the right is oriented toward the fixed point. This point is therefore a half-stable fixed point because any time course with initial conditions of $[Protein]_0 > 0$ will approach a steady state of $[Protein] = 0$; initial conditions of $[Protein]_0 < 0$ lead to rapidly decreasing protein concentrations.

A most interesting conclusion can therefore be drawn from our analysis: The number of fixed points, a critical determinant of system dynamics, depends on the value of $k_{trl,max}$! Figure 4.9a summarizes the vector fields that we produced in Figures 4.6–4.8. When $k_{trl,max}$ is large and positive, the fixed points, one unstable and one stable, are located at a certain distance that becomes increasingly smaller as $k_{trl,max}$ approaches zero. At $k_{trl,max} = 0$, the two fixed points fuse into one half-stable fixed point, which disappears at negative values of $k_{trl,max}$.

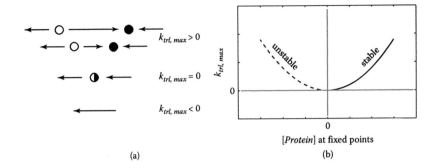

FIGURE 4.9   Summary of the observations in Practice Problem 4.2. (a) Schematic of the changes in the vector field for different values of $k_{trl,\,max}$. Stable fixed points appear as filled circles, unstable fixed points are empty circles, and the half-stable fixed point is half shaded. (b) Using a bifurcation diagram to visualize the effect of $k_{trl,\,max}$ on protein concentration at the fixed points.

Figure 4.9b (plotted from Equation 4.6) depicts the relationship between $k_{trl,\,max}$ and the fixed points in more detail. A change in the number of fixed points, caused by a change in parameter values, is called a **bifurcation**, and the plot in Figure 4.9b is called a **bifurcation diagram**. There are many types of bifurcations; this case illustrates a **saddle-node bifurcation**.

With regard to relevance, it is biologically impossible to have a $k_{trl,\,max}$ less than zero. But sometimes, for example in the case of the inducible promoter that we considered in Chapter 3, transcription can be halted, and therefore $k_{trl,\,max} = 0$.

To summarize what we have seen so far in this chapter, graphical analysis enables us to quickly visualize the results of changing initial conditions or parameters, much like our previous Boolean analysis, but in more detail.

ADDING FEEDBACK

Now that we have addressed the system without feedback described in Equation 4.1, let's move to the system with feedback shown in Equation 4.2. The protein production term in this equation is substantially more complicated than in Equation 4.1, making the entire equation somewhat more difficult to plot without the aid of a computational tool.

Fortunately, there is an easier way to generate our vector field: Instead of plotting the complete equation for $d[Protein]/dt$, we can plot each term (the production rate v_{prod} and the loss rate v_{loss}) on the right side of the equation separately and compare those two plots.

Comparing the magnitude of the protein production (Figure 4.10a, gray) and loss (Figure 4.10a, black) rates yields the vector field (Figure 4.10b). At low protein concentrations, protein production is substantially greater than protein loss, so we draw an arrow toward higher protein concentrations (pointing to the right). Moving rightward, the difference between the production and loss rates shrinks; thus, the arrows pointing right are shorter. High protein concentrations lead to a loss rate that is higher than the production rate, so the vectors at high concentrations point to the left. Finally, there is a single point at which the production rate is equal to the loss rate; therefore, $d[Protein]/dt = 0$. This point is fixed, and because the vectors in the field all point toward this point, we know that it is a stable fixed point.

Let's think about whether changing some of our parameters would lead to a bifurcation. Figure 4.11 illustrates the outcome of changing k_{ploss}, $k_{trl, max}$, and K individually. In each case, the protein production and loss rates only meet at a single point, and as a result, changing these parameters will not lead to a bifurcation. In order for these rates to meet at multiple points, one or the other terms would need significantly more curvature, or nonlinearity. For example, cooperativity introduces nonlinearity into a system, which therefore may lead to a bifurcation.

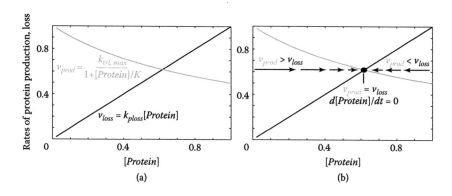

FIGURE 4.10 Graphing the dynamics of the system (with feedback) in Equation 4.2. (a) Protein production (gray) and loss (black) in the system. (b) The vector field reveals a single stable fixed point.

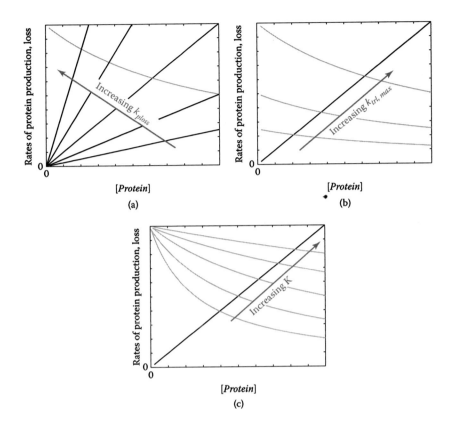

FIGURE 4.11 The effects of changing k_{ploss} (a), $k_{trl,\,max}$ (b), or K (c) in Equation 4.2 on the position of the fixed point. Protein production is shown in gray, and protein loss appears in black. The point at which the gray and black lines intersect is the fixed point of the system. Notice that none of these changes leads to a bifurcation; there is only one fixed point in each of the intersections shown.

TWO-EQUATION SYSTEMS

To this point, our focus has been on graphical analysis of single equations. However, our original system depended on two equations, one for protein, and another for mRNA:

$$\frac{d[Protein]}{dt} = k_{trl}[mRNA] - k_{ploss}[Protein] \tag{4.7}$$

$$\frac{d[mRNA]}{dt} = k_{trs,\,max}\left(\frac{1}{1+[Protein]/K}\right) - k_{mloss}[mRNA] \tag{4.8}$$

In Chapter 3, we assumed that the mRNA concentration reaches a steady state much more quickly than the protein concentration, which enabled us to solve Equation 4.8 for $d[mRNA]_{ss}/dt = 0$ and then to substitute the resulting value of $[mRNA]_{ss}$ into Equation 4.7 to obtain Equation 4.2. In this case, we want to solve both equations together without making any additional assumptions; graphical analysis will be helpful.

We have two variables to worry about instead of one, but otherwise, our approach will be similar to our approach for single equations. We start by considering all the steady states together and drawing **nullclines**, which are the steady-state solutions of each equation in the system. As usual, we set $d[Protein]/dt$ and $d[mRNA]/dt$ equal to zero and solve for $mRNA$ in terms of $Protein$:

$$[mRNA]_{ss} = \frac{k_{ploss}}{k_{trl}}[Protein]_{ss} \tag{4.9}$$

$$[mRNA]_{ss} = \frac{k_{trs}}{k_{mloss}}\left(\frac{1}{1+[Protein]_{ss}/K}\right) \tag{4.10}$$

Figure 4.12 contains plots of both of these lines for $K = k_{ploss}/k_{trl} = k_{trs}/k_{mloss} = 1$. This plot looks a lot like Figure 4.10, and for good reason, since they depict the same equations. This should not surprise you because our second system is a more complicated version of our first system.

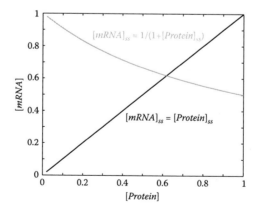

FIGURE 4.12 Including the change in mRNA production in our analysis of the system with feedback. In this two-dimensional case, the black and gray lines are nullclines. Also, notice the similarity to Figure 4.10a.

Notice that there is only one point of intersection between the nullclines, and we identify those coordinates (0.62, 0.62) by solving Equations 4.9 and 4.10. This location reflects the point at which the protein and mRNA levels, and therefore the system as a whole, are at a steady state.

Again, we draw the vector field to determine how the system approaches that steady state. Let's start with four individual vectors at each corner of the plot in Figure 4.12; the first vector will be the origin (0, 0). If we substitute these values for *Protein* and *mRNA* into Equations 4.7 and 4.8, we obtain:

$$\frac{d[Protein]}{dt} = [mRNA] - [Protein] = 0 - 0 = 0 \tag{4.11}$$

$$\frac{d[mRNA]}{dt} = \left(\frac{1}{1+[Protein]}\right) - [mRNA] = 1 - 0 = 1 \tag{4.12}$$

As a result, we draw a vector that points upward in the *y* direction. Using the same approach, we identify the three other corners of the plot in Figure 4.13a and draw the four vectors in Figure 4.13b.

You see that it is fairly easy to calculate what d[*Protein*]/dt and d[*mRNA*]/dt should be, given any initial conditions. Even with this small

	$\frac{d[Protein]}{dt}$	$\frac{d[mRNA]}{dt}$
P = 0, M = 0	0	1
P = 0, M = 1	1	0
P = 1, M = 1	0	-0.5
P = 1, M = 0	-1	0.5

(a)

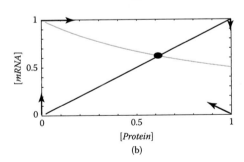

(b)

FIGURE 4.13 Computing (a) and visualizing (b) the vector field for our system with feedback. Note that the outside vectors do not point directly at the fixed point. The vector lengths shown here are scaled down 1:4 to facilitate visualization. P, [*Protein*]; M, [*mRNA*].

beginning, you may notice a few things, for example that these outside vectors do not point directly at the fixed point, and that there are times when only the protein concentration or only the mRNA concentration changes as the immediate response to a set of initial conditions.

A computational package such as MATLAB makes these calculations even easier, and you will learn how to perform some of these calculations in the problems at the end of the chapter. A much more complete vector field, generated using MATLAB, appears in Figure 4.14a.

Notice that the vectors seem to spiral toward the fixed point, allowing us to conclude that the fixed point is stable. For further evidence, we can plot specific trajectories on the vector field using the same four initial conditions

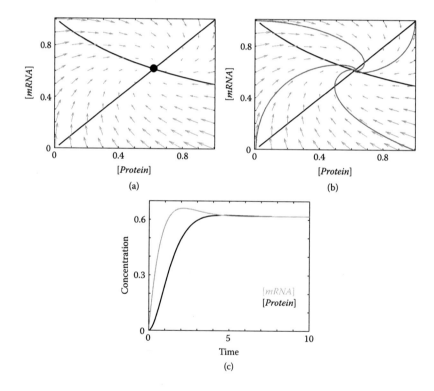

FIGURE 4.14 Visualizing the changes in mRNA and protein concentration in our system with feedback. (a) The complete vector field. (b) Plotting trajectories (red) for the four initial conditions in Figure 4.13 demonstrates that the system moves toward the stable fixed point. (c) The dynamic responses of the mRNA (gray) and protein (black) concentrations over time beginning at conditions (0, 0). Notice that [mRNA] overshoots the stable state. The MATLAB functions meshgrid and quiver were used to create the vector fields in (a) and (b).

that we considered in Figure 4.13b, for example. These trajectories (also calculated by solving the ODEs numerically in MATLAB) indeed all move toward the stable fixed point in Figure 4.14b.

However, they do not move directly; all four trajectories are curved in Figure 4.14b. In some cases, the trajectory even passes the fixed-point value of either [mRNA] or [Protein] before reaching a steady state at both values.

These interesting aspects of the trajectories carry over to the time courses. Figure 4.14c depicts the dynamic responses of [mRNA] and [Protein] given initial conditions with both concentrations equal to zero. The corresponding red trajectory in Figure 4.14b begins at the origin and moves up and to the right before reaching the stable fixed point. The mRNA concentration rises quickly, overshoots the steady-state value, and then returns to it; the protein concentration undergoes a more simple rise to the steady-state value.

Other trajectories lead to different time courses, but you can see from the vector field that overshoot commonly occurs by one variable or the other. These time courses were drawn using MATLAB, but the real key to drawing them is the vector field.

I hope you can see that graphical analysis has several nice features. First, we exploit the specificity and detail that accompany ODEs, but like the Boolean approach, we are able to examine many responses simultaneously to obtain an idea of how the system works in general. We can also relax some of the assumptions required to make analytical solutions of ODEs tractable; note, for example, that the results in Figure 4.14c are inconsistent with our previous assumption (made largely for reasons of expediency) that the mRNA concentration reaches a steady state *before* the protein concentration. Finally, we can see how parameter changes affect system dynamics.

The challenge with these approaches is that it is somewhat difficult to visualize information in multidimensional space. As a result, the graphical approach is powerful for two- or three-dimensional systems, but not for higher systems. If our system is small enough, however, or if we can make a realistic reduction in the complexity of our system, it is hard to beat graphical approaches for the intuition they can give us.

CHAPTER SUMMARY

Graphical analysis can lead us to deeper intuition about ODEs by enabling us to identify stable and unstable states, to plot dynamics, and to consider the effects of varying initial conditions and parameter values, rapidly and without any need for a computer. We began with a single ODE system, plotting the values of a given variable (some variable X) against the change in that variable over time (dX/dt). The fixed points of the system are defined as the x intercepts of the plot of X versus dX/dt. The stability of the fixed points was determined by calculating a one-dimensional vector field in which the direction of the vector was in the positive direction when $dX/dt > 0$ and in the negative direction when $dX/dt < 0$. The magnitude of the vector is simply the absolute value of dX/dt (although these vectors were scaled down in the figures to make them easier to visualize).

The vector field indicates in which direction and at what speed a given value of X will change. Stable fixed points are those in which the vectors on both sides point toward them, and unstable fixed points are surrounded by vectors pointing away from them; there are also half-stable points that are stable on one side and unstable on the other. We can use the dX/dt plots, together with the fixed points and vector field, to quickly sketch the time-course plots for X for any initial conditions.

We also discussed how the fixed points and vector field change for different ODEs; in particular, we saw that nonlinearity was required for a single ODE to have multiple fixed points. Changing the values of parameters in an ODE can also lead to changes in the fixed points and vector field; the study of these changes, in particular when parameter value changes lead to loss or gain of a fixed point, is called bifurcation analysis.

Graphical analysis can be adapted for more complicated single-ODE systems by plotting the positive terms in the ODE separately from the negative terms and using the difference between these two plots to create the vector field. This analysis can be expanded to two-ODE systems. The nullclines of the system, one line for each ODE in which dX/dt is set to zero and the result is plotted, are plotted together with a two-dimensional vector field. The fixed points in this case are where the nullclines intersect.

The main weakness of graphical analysis is that it can only be applied to small systems. However, if you have or can obtain a relatively small (one- or two-dimensional) model of your circuit, it can be a powerful way to build understanding.

RECOMMENDED READING

Ellner, S. and Guckenheimer, J. *Dynamic Models in Biology*. Princeton, NJ: Princeton University Press, 2006.

Gardner, T. S., Cantor, C. R., and Collins, J. J. Construction of a genetic toggle switch in *Escherichia coli*. *Nature* 2000, 403(6767): 339–342.

Huang, D., Holtz, W. J., and Maharbiz, M. M. A genetic bistable switch utilizing nonlinear protein degradation. *Journal of Biological Engineering* 2012, 6:9–22.

Strogatz, S. H. *Nonlinear Dynamics and Chaos: With Applications to Physics, Biology, Chemistry, and Engineering (Studies in Nonlinearity)*. Boulder, CO: Westview Press, 2001.

PROBLEMS

PROBLEM 4.1
A Simple Oscillatory System

This problem is more illustrative than biological in order to build your skills. Consider a protein whose concentration over time can be described with the following ODE:

$$\frac{dp}{dt} = a \cdot p - \cos(p)$$

where a is a constant equal to 0.25, p is the protein concentration, and t is time. As you learned, we can plot the two terms on the right side of this equation separately to determine fixed points and to generate a vector field. In other words, we can plot $f(p)$ versus protein concentration, where:

$$f(p) = \begin{Bmatrix} 0.25 \cdot p \\ \cos(p) \end{Bmatrix}$$

The resulting plot shows the upper term in gray and the lower term in black.

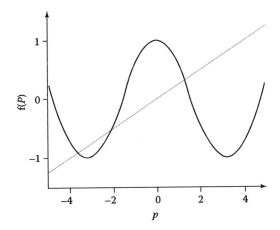

a. By hand, draw the vector field describing the change in protein concentration. Make sure to note both direction and magnitude in your vector field.

b. Circle all of the steady states, if applicable. Fill in the steady state(s) that are stable.

c. By hand, sketch the changes in protein concentration over time for at least five initial concentrations of protein, including 3.5, 1.0, −1.8, −3.2, and −4.5.

d. What happens to the fixed points as you increase the value of a? You may use graphs to illustrate your answer.

PROBLEM 4.2
An Autoregulatory Circuit with Positive and Cooperative Feedback

In Chapter 3, we considered circuits that contain positive feedback. We also considered the effects of cooperativity, for example when multiple copies of the protein were required to interact at the DNA promoter for the feedback to be effective. Here, we consider positive and cooperative feedback together, as shown in the figure below.

In this case, a protein dimer (Hill coefficient $H = 2$) is required to bind to DNA for transcription to occur. The ODEs describing this system are:

$$\frac{d[Protein]}{dt} = k_{trl} \cdot [mRNA] - k_{ploss} \cdot [Protein]$$

$$\frac{d[mRNA]}{dt} = \frac{k_{trs,max} \cdot [Protein]^2}{\left(\left(\frac{k_d}{k_a}\right)^2 + [Protein]^2\right)} - k_{mloss} \cdot [mRNA]$$

For the following questions, let $k_{trl} = k_{trs,max} = k_d = k_a = 1$; $k_{ploss} = 0.02$; and $k_{mloss} = 20$. Don't worry about units for now (they've got plenty of time to break your heart).

a. Use MATLAB to plot the nullclines of the equations in this problem. Hints: Let mRNA be your y axis and protein be your x axis; let your protein concentration range from 0 to 3.

b. How many fixed points does this system have? Circle them on your graph from (a).

c. Using your graph from (a), draw the vector field for several points of your choice by hand. Which of the fixed points are stable?

d. Explain what the stable points mean in terms of the biology of the system. How can changes in the parameter k_{ploss} lead to different numbers of fixed points? Provide at least two examples.

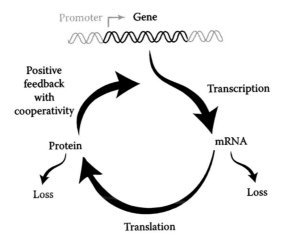

PROBLEM 4.3
A Single Feedback Loop System with Enzymatic Degradation

We have mostly focused on production terms in our ODEs when considering the regulation of gene expression. However, protein degradation rates are also important in regulatory circuits and can be critical for the regulation of bistable networks. In this problem, we look at the use of dynamic protein degradation to create a synthetic bistable switch, using the Lon protease found in *Mesoplasma florum* (*mf*-Lon).

In the hybrid switch (described in Huang, Holtz, and Maharbiz, 2012), the positive autoregulatory loop is identical to the one found in lambda phage, where promoter P_{RM} expresses the lambda repressor cI, which is also a transcriptional activator of P_{RM}. Bistable behavior may be possible using only this single positive-feedback loop along with constitutive expression of *mf*-Lon, which quickly degrades cI. However, the Michaelis constant for *mf*-Lon would need to be significantly smaller than the value reported by Huang et al. for bistable behavior to occur, as you will see. We will analyze and model this simplified circuit in three ways to uncover its important dynamic properties.

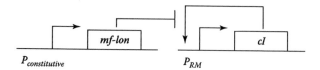

The equations that govern the system are:

$$\frac{d[m_{cI}]}{dt} = k_{trs,\,max}\left(\cfrac{1}{1+\left(\cfrac{K_{Diss}}{[p_{cI}]}\right)^{H}} + l_{PRM} \right) - k_{mloss}[m_{cI}]$$

$$\frac{d[p_{cI}]}{dt} = k_{trl}[m_{cI}] - k_{ploss}[p_{cI}] - k_{cat}[p_L]\left(\cfrac{1}{1+\cfrac{K_M}{[p_{cI}]}} \right)$$

where H equals the Hill coefficient of the system, K_{Diss} is the dissociation constant of cI from P_{RM}, l_{PRM} is the fraction of mRNA transcription that is caused by leakage from the P_{RM} promoter (the basal expression level), k_{cat} is the catalytic rate of mf-Lon, p_L is the protein concentration of mf-Lon, and K_M is the Michaelis constant for mf-Lon (Michaelis constants are discussed in more detail in Chapter 9). As described in Problem 3.4, when multiple repressor binding sites occur in front of a gene, the Hill coefficient describes the cooperative binding properties of that system (for two binding sites, $H = 2$, etc.).

To start, let's set:

$$k_{trs,\,max} = 1.35 \times 10^{-9}\ \text{M}^2\ \text{s}^{-1}$$

$$K_{diss} = 2.5 \times 10^{-8}\ \text{M}$$

$$H = 1$$

$$l_{PRM} = 0.1$$

$$k_{mloss} = 2.38 \times 10^{-3}\ \text{s}^{-1}$$

$$k_{trl} = 5 \times 10^{-5}\ \text{s}^{-1}$$

$$k_{ploss} = 2 \times 10^{-4} \text{ s}^{-1}$$

$$k_{cat} = 0.071 \text{ s}^{-1}$$

$$p_L = 1 \times 10^{-10} \text{ M}$$

$$K_M = 3.7 \times 10^{-9} \text{ M}$$

a. List at least two of the assumptions we've made to produce the two equations that describe the system.

b. In Problem 3.4 in Chapter 3, the protein production rate was proportional to:

$$\frac{1}{1+\left(\dfrac{[Protein]}{K_{Diss}}\right)^{H}}$$

The production term in our ODE for m_{cI} is different in that K_{Diss} is in the numerator and p_{cI} is in the denominator. Where do these changes come from? Show mathematically how this term was derived.

c. What does K_M mean in biochemical terms in this system? Explain why each of the variables appears the way it does in the term describing the enzymatic degradation of cI by *mf*-Lon.

d. Plot the nullclines of these two equations in MATLAB. Circle any fixed points.

e. Plot the vector field using the MATLAB commands `quiver` and `meshgrid`. Are the fixed points you identified stable or unstable?

f. So far, we have assumed that $H = 1$. In fact, the P_{RM} promoter has two adjacent operator sites; binding of one cI protein to the upstream "helper" operator helps recruit another cI protein to the "activator" operator, which when bound stimulates transcription. Thus, positive cooperativity occurs in this system, and $H = 2$ for P_{RM}. Plot the new nullclines in MATLAB and circle the fixed points for this scenario.

g. Plot a vector field on your graph from (f) and determine which of the fixed points are stable. What do these fixed points mean for the biology of the system?

h. Now, vary the value of K_M. How does varying the value of K_M change the stability of the system? Use the MATLAB subplot command to add these new plots to one figure to obtain a single figure that allows visual inspection of the effect of changing K_M.

PROBLEM 4.4
A Dual-Repressor System

Here we return to the toggle switch from Problem 2.2 (Gardner, Cantor, and Collins, 2000), which is diagrammed below.

The equations that govern the system are:

$$\frac{d[Repressor_1]}{dt} = k_{trl} \cdot [mRNA_1] - k_{ploss} \cdot [Repressor_1]$$

$$\frac{d[mRNA_1]}{dt} = \frac{k_{trs,\,max1}}{\left(\left(\frac{[Repressor_2]}{K_{Diss2}^2}\right)^H + 1\right)} - k_{mloss} \cdot [mRNA_1]$$

$$\frac{d[Repressor_2]}{dt} = k_{trl} \cdot [mRNA_2] - k_{p,\,loss} \cdot [Repressor_2]$$

$$\frac{d[mRNA_2]}{dt} = \frac{k_{trs,\,max2}}{\left(\left(\frac{[Repressor_1]}{K_{Diss1}^2}\right)^H + 1\right)} - k_{mloss} \cdot [mRNA_2]$$

where $k_{ploss} = k_{mloss} = k_{trl} = 1/s$; $k_{trs,\,max1} = 3/s$; $k_{trs,\,max2} = 4/s$; $K_{Diss1} = 0.6\ \mu M$; $K_{Diss2} = 0.8\ \mu M$; and $H = 1$.

a. For both repressors, assume that $[mRNA]$ quickly reaches a steady state with respect to $[Protein]$. Use this assumption to derive a new, reduced set of equations for the change in protein concentrations over time.

b. In MATLAB, plot the nullclines of the two equations you found in (a).

c. Plot a vector field on the graph for (b) and use it to determine the stability of any fixed points you identified.

d. Now let's analyze what would happen if we added an additional DNA binding site to the promoter for each repressor, such that $H = 2$. Plot the new nullclines in MATLAB and circle any fixed points.

e. Draw a vector field on the graph from (d) and determine which of the fixed points are stable.

f. Explain what these fixed points mean for the biology of the system.

g. Now vary K_{Diss} for one of the repressors. How does this change affect the stability of the system?

CHAPTER 5

Variation: Numerical Integration

- Calculate the numerical solutions to sets of ODEs using the Euler, midpoint, and Runge–Kutta methods

- Approximate the error of these methods using a Taylor series expansion

- Understand the sensitivity of each method to the time-step size and how this sensitivity relates to the method order

We conclude our exploration of methods for solving ODEs by discussing numerical methods. Even a few years ago, I would have been tempted to cover this entire chapter with two short sentences:

Open MATLAB on your computer.

Type "help ode45" at the command line.

In other words, the methods I am about to describe have already been implemented in MATLAB, so why do you need to learn them? The first reason is that you will be better able to use the MATLAB functions if you really understand how they work and what their limitations are. That is not the most important reason, however. I want you to learn these methods so that you can "roll your own" versions of the methods, adapting them to your own needs as necessary. In particular, your ability to alter and, when necessary, subvert these numerical solving techniques will be critical to creating **hybrid models** that combine ODEs with non-ODE-type methods. I come back to this point in Chapter 10.

THE EULER METHOD

Numerical integration of ODEs depends on a fairly straightforward problem: Given a function $f(x) = dx/dt$ and values for x and t at a particular instant (I'll call them x_0 and t_0, respectively), determine the value of x at a later time point (x_1 and t_1). If you recognize that $dx/dt = f(x)$ is the form of an ODE, then you can probably also see that our protein and mRNA equations from Chapters 3 and 4 can be solved using these same approaches.

Not to drag this book back into the gutter, but let's revisit the analogy I developed in Chapter 3 (Figure 5.1, a refresher, was originally Figure 3.2). Remember, I was estimating the amount of water in my rain gutter over time by measuring the rates of rainfall and drainage using this mass balance equation:

$$GW(t + \Delta t) = GW(t) + Rain(t) \cdot \Delta t - Drain(t) \cdot \Delta t \qquad (5.1)$$

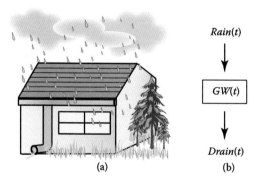

(a) (b)

FIGURE 5.1 Conceptualizing ODEs with a rain gutter analogy (See also Figure 3.2). (a) During a rainstorm, water enters the gutter at rate *Rain*, flows out of the drainpipe at rate *Drain*, or builds up in the gutter to an amount *GW*. (b) The compartment diagram for the analogy indicates that water only enters the gutter from the rain and only exits the gutter from the drain; the diagram also captures the dependence on time of these processes.

I then derived an ODE from this equation:

$$\frac{dGW}{dt} = Rain(t) - Drain(t) \tag{5.2}$$

We can cast the central problem of this chapter in terms of this rain gutter equation by recognizing that GW is analogous to x, $Rain(t) - Drain(t)$ to $f(x)$, t in Equation 5.1 to t_0, and $(t + \Delta t)$ to t_1. In that context, let's rearrange Equation 5.1:

$$GW(t + \Delta t) = GW(t) + (Rain(t) - Drain(t)) \cdot \Delta t \tag{5.3}$$

In other words, we can numerically predict the amount of water in the gutter at a future time if we know the present amount of water, the present rates of rainfall and drainage, and the interval between the present and future time. Now, let us apply our analogy to approximate x_1 based on what we know about x_0:

$$x_1 = x_0 + f(x_0) \cdot \Delta t \tag{5.4}$$

To get a feeling for what Equation 5.4 really means, take a look at Figure 5.2. To calculate x_1 using Equation (5.4), you draw a line extending from, and tangent to, the point (x_0, t_0). You then approximate x_1 as the

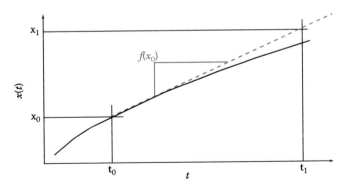

FIGURE 5.2 The Euler method. By drawing a line tangent to our curve (red dashed line) at point (x_0, t_0), we can approximate the value of point (x_1, t_1) from the slope of the tangent line. This approximation becomes less valid as the curve diverges from the tangent, so we want to take the smallest step possible between t_0 and t_1 to generate the most accurate value for x_1. Note that when $t_0 = 0$, Equation 5.4 is the equation of a line with intercept x_0 and slope $f(x_0)$.

value of x where the tangent line intersects with $t = t_1$. In fact, Equation 5.4 is the equation of a line, and if t_0 is set as the origin, x_0 is the intercept and $f(x_0)$ is the slope.

The method that outlined here is called the **Euler method**. You should note two things about applying the Euler method. First, you can see from Figure 5.2 that the accuracy of the approximation depends on the magnitude of Δt. When t_1 is far enough from t_0 (in Figure 5.2, about halfway between t_0 and t_1), the approximation starts to diverge significantly from the actual solution. The second thing to notice is that our function f is evaluated at x_0 when we use the Euler method. I elaborate on how to choose a useful time step, as well as other ways to evaluate f, in further discussion.

Now that we have established how to implement the Euler method for a single time step, it will not be too difficult to describe the numerical solution for a large number of time steps. I will use a MATLAB-esque pseudocode here:

```
(1) Define simulation timespan, starting at (t0, x0),
    and divide the timespan into N steps such that
    deltaT = timespan/N is sufficiently small.
(2) Define two arrays to hold the time and
    concentration data: tarr = [t0]; xarr = [x0];
(3) for i = 1:N
            tnew = tarr(i) + deltaT;
            xnew = xarr(i) + f(xarr(i)) * deltaT;
            tarr = [tarr tnew];
            xarr = [xarr xnew];
    end
(4) plot(tarr, xarr)
```

What it means to be "sufficiently small" is addressed in the next section.

ACCURACY AND ERROR

Let's apply the Euler method to a familiar example to explore how it works. Revisit the equation most recently seen as Equation 4.1 (and previously in Equations 3.17 and 3.18): the simple expression of protein without feedback. The two kinetics constants in this equation govern the rate of transcription and the rate of decay. To make the analysis relatively straightforward, let's set both of these rates equal to 1, obtaining the following equation:

$$\frac{d[Protein]}{dt} = 1 - [Protein] \tag{5.5}$$

We determined the general solution to this equation analytically in Equation 3.25; with the constants equal to 1 and initial conditions of (0,0), this solution simplifies to:

$$[Protein](t) = 1 - e^{-t} \qquad (5.6)$$

The analytical solution to Equation 5.6 is plotted in Figure 5.3. Now let's try using the Euler approximation. We start with the same initial conditions. Our function $f(Protein)$ is determined from Equation 5.5: $f(Protein) = 1 - Protein$. If we try a time step of 0.01, we can calculate the value of *Protein* at time = 0.01 as:

$$[Protein]_{(t=0.01)} = [Protein]_0 + 0.01 \cdot (1 - [Protein]_0) \qquad (5.7)$$

which evaluates to $0 + 0.01 \cdot 1 = 0.01$. Interestingly, if we substitute $t = 0.01$ into Equation 5.6, we also obtain ~0.01. So, for the first time step at least, the Euler method provides a good approximation. We follow the same procedure to determine the rest of the solution, and it turns out that the line from the Euler method completely overlaps the line from the analytical solution (Figure 5.4).

So far, so good: Apparently, a time step of 0.01 is sufficiently small to give us a good approximation in this case. How did I know to choose 0.01 as the time step? I just tried a few possibilities, and the fit that I saw

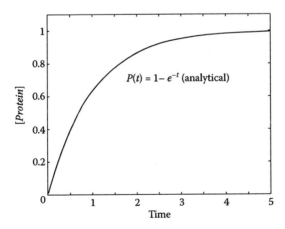

FIGURE 5.3 The analytical solution for our simple protein system. The ODE (Equation 5.6) describes simple protein expression and decay, with the rate constants of both set to 1, without feedback.

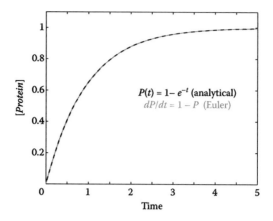

FIGURE 5.4 The Euler method for $t = 0.01$ provides an accurate approximation for Equation 5.6. Here, the line we generated using the Euler method (dashed gray) completely overlaps the line from the analytical solution (Figure 5.3; black line). A time step of $\Delta t = 0.01$ is therefore sufficiently small to return a reasonable approximation via the Euler method.

in Figure 5.4 looked good to me! I can also show you what some of my other attempts looked like. For example, what happens when we move to a bigger time step? The answer appears in Figure 5.5, which plots the Euler approximation of Equation 5.5 for increasingly large time steps. Notice that the approximation quickly becomes inaccurate as the time step is increased.

It would be nice to dig a little deeper with this observation, especially to examine ways to describe the accuracy of a given method and time step. Two ways to look at the error are discussed; the first is specific to our example, and the second is more generalizable.

First, let's look specifically at our example. For this case, we can define an error by determining the **Euclidean distance** between the analytical solution and the Euler approximation at time = 1, 2, 3, 4, and 5 as follows:

$$error = \sum_{i=1}^{5} \left([Protein]_{analytical,\ t=i} - [Protein]_{Euler,\ t=i} \right)^2. \qquad (5.8)$$

Table 5.1 shows the error (Equation 5.8) for four time step lengths. As you saw in Figure 5.5, the error increases substantially with time step length, and the increase is nonlinear: As the time step increases from 0.01 to 0.1, the error increases by 10 fold, but with the increase from 0.1 to 1.0, the error increases by 15 fold.

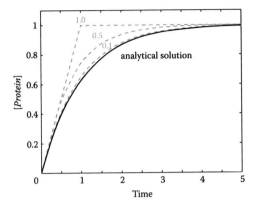

FIGURE 5.5   Increasing the size of the time step generates less accurate approximations via the Euler method. Time steps of $\Delta t = 0.1$, 0.5, and 1.0 (dashed gray lines) lead to larger and larger deviations from the curve of the analytical solution to Equation 5.6 (black line). Recall that a time step of $\Delta t = 0.01$ did not deviate from the analytical solution in Figure 5.4.

TABLE 5.1   Error Introduced by the Euler Approximation at Various Time Steps, Calculated Using Equation 5.8

Time Step	Error
0.01	0.0045
0.1	0.0455
0.5	0.2450
1.0	0.5781

Equation 5.8 is admittedly a bit artificial (we cannot apply it to compare numerical techniques for every equation we want to solve), so we will also use a more general approach for determining error. One way involves the use of a **Taylor series approximation**. We will not discuss the details of the approach here, but briefly, if you have a function S defined in terms of some variable y and want to approximate the value of $S(y)$, when y is near a value a for which $S(a)$ is known, the Taylor series approximation is:

$$S(y) \approx S(a) + \frac{S'(a)}{1!}(y-a) + \frac{S''(a)}{2!}(y-a)^2$$

$$+ \frac{S'''(a)}{3!}(y-a)^3 + ... + \frac{S^n(a)}{n!}(y-a)^n$$

(5.9)

where $S'(a) = dS/dy$, evaluated at a. For this approximation to work, S must be "infinitely differentiable" (S', S'', and so on), even though in practice we only use the first handful of terms in the approximation to simplify the calculation.

The Taylor series approximation gives us a second, independent numerical method (the first method in this case is the Euler method itself) to predict future values of a function. Another way to calculate error, particularly when the analytical solution is not known, would therefore be to compare approximations from a variety of methods. For example, we could approximate the "error" of the Euler method as the difference between values calculated using it and the Taylor series approximation.

Referring to Equation 5.4, recall that we determined a new value of x (x_1) in terms of a previous value x_0, a function that described the time dynamics $f(x)$ and the time step Δt. If we substitute $x(t)$ for $S(y)$, let $t_1 = t_0 + \Delta t$, and set $a = t_0$, Equation 5.9 becomes:

$$x(t_1)_{Taylor} \approx x(t_0) + \frac{1}{1!} \frac{dx}{dt}\bigg|_{x(t_0)} (t_0 + \Delta t - t_0) + \frac{1}{2!} \frac{d^2x}{dt^2}\bigg|_{x(t_0)} \quad (5.10)$$

$$(t_0 + \Delta t - t_0)^2 + \ldots$$

Simplification leads to:

$$x(t)_{Taylor} \approx x_0 + \Delta t \cdot f(x_0) + \tfrac{1}{2} \Delta t^2 \cdot f'(x_0) + \ldots \quad (5.11)$$

Now let's define the error of the Euler method simply as the absolute difference between the values of x_1 determined using the Euler method and using the Taylor series approximation:

$$error_{Euler} = \left| x_{1,Taylor} - x_{1,Euler} \right| \quad (5.12)$$

Substituting Equations 5.4 and 5.11 and ignoring terms past the second derivative in Equation 5.11 (multiplying these terms by higher powers of Δt makes these terms relatively small), we obtain:

$$error_{Euler} = \tfrac{1}{2} f'(x_0) \Delta t^2 \quad (5.13)$$

Note that the $f'(x_0)$ term is a constant because it is evaluated at a specific value of x (x_0). Equation 5.13 therefore reduces to:

$$error_{Euler} \propto \Delta t^2 \qquad (5.14)$$

In plain English, the error depends on the square of the time step length, as we noticed in Figure 5.5 and Table 5.1. The Euler method is called "first order" (the **order** of the method is the power of Δt in Equation 5.14 minus 1) because of this dependence.

THE MIDPOINT METHOD

The Euler method is useful for teaching the concepts of numerical integration, but with real-world problems like ours for which error is a concern, we will need higher-order methods. "Higher order" means that the error depends on a higher power of Δt, so higher-order methods are generally more accurate at a given Δt. Let's consider a second-order approach called the midpoint method (Figure 5.6).

The left panel of Figure 5.6 is an illustration of the Euler method that is similar to Figure 5.2 except the curve is a little more dramatic so that you can see the error more clearly. As you can see, what's causing the error in the Euler method is that big slope, which in the example is larger at x_0 than at any later time point. If we approximated x_1 using a smaller slope

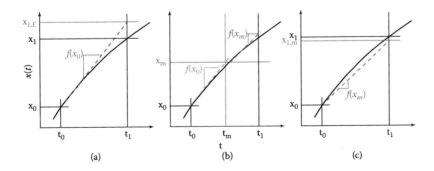

FIGURE 5.6   Illustration of the midpoint method to approximate the value of x_1. Left, application of the Euler method leads to an assignment of $x_{1,E}$ based on the slope of the tangent to $f(x)$ at x_0, denoted as $f(x_0)$. The error of the Euler method versus the analytical solution is $x_1 - x_{1,E}$, which is large here. Middle, use of the Euler method to approximate x_m (Equation 5.16). Right, solving Equation 5.15 with x_m and $f(x_m)$ yields a value of x_1 that is too small, but with smaller error than that obtained by Euler's method alone.

determined from elsewhere in the curve, it could yield a more accurate calculation.

The midpoint method is simple in principle: Determine the slope at the midpoint and use that slope to calculate x_1. First, we modify Equation 5.4 to include the midpoint:

$$x_{1,m} = x_0 + f(x_m) \cdot \Delta t \tag{5.15}$$

where $x_{1,m}$ is the approximation of x_1 calculated using the midpoint method, and x_m is the value of x halfway across the time interval (at time t_m, equal to $(t_0 + t_1)/2$).

Wait a minute, you're thinking (I hope)—if we do not know x_1, how do we know x_m? Well, we don't know x_m! We can only approximate it, and the tool that we have is ... the Euler method! x_m is therefore calculated as:

$$x_m = x_0 + f(x_0) \cdot \frac{\Delta t}{2} \tag{5.16}$$

The calculation of x_m is also shown in the middle panel of Figure 5.6. Notice that the Euler method predicts a value for x_m that is also too large. As a result, $f(x_m)$ will not actually be the slope of the curve at t_m but at a higher value of t, which in this case has a smaller slope. Once x_m and $f(x_m)$ are determined, they can be used to solve Equation 5.15 to yield $x_{1,m}$ (Figure 5.6, right). Alternatively, Equations 5.15 and 5.16 can be combined into a single equation:

$$x_{1,m} = x_0 + f\left(x_0 + \tfrac{1}{2} f(x_0)\Delta t\right) \cdot \Delta t \tag{5.17}$$

Figure 5.7 illustrates how the midpoint and Euler methods compare over a larger span of the curve from the analytical solution. Note that in our case, the Euler method overpredicts x_1, while the midpoint method underpredicts it. This effect is caused by the difference in slope discussed earlier. Using the Taylor series approximation, we can estimate the error of the midpoint method as before. The error turns out to be proportional to Δt^3, so the midpoint method is second order. For Figure 5.7, the error (Equation 5.8) is 0.3907 for the midpoint method, which is 60% of the Euler method error at $\Delta t = 1$.

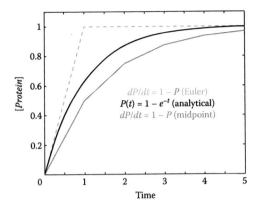

FIGURE 5.7 The midpoint method (dark gray) versus the Euler method (dashed light gray) and the analytical solution (black) for $\Delta t = 1$. (meaning that only the points {1, 2, 3, 4, 5} were calculated using the midpoint and Euler methods; the lines connecting the points were added by MATLAB).

PRACTICE PROBLEM 5.1

Given the equation:

$$\frac{dp}{dt} = \frac{p^3 + 2p}{3}$$

and that $p(0) = 2$, $t(0) = 0$, and $\Delta t = 0.5$, use the midpoint method to find $p(t = 0.5)$. Show your work.

SOLUTION

First we find p_m using the Euler method and Equation 5.16:

$$p_m = p_0 + f(p_0) \cdot \frac{\Delta t}{2}$$

$$p_m = 2 + \left(\frac{2^3 + 2 \cdot 2}{3} \right) \cdot \frac{0.5}{2} = 3$$

Next, we substitute the value for p_m into Equation 5.15:

$$p_{1,m} = p_0 + f(p_m) \cdot \Delta t$$

$$p_{1,m} = 2 + \left(\frac{3^3 + 2 \cdot 3}{3} \right) \cdot 0.5 = 7.5$$

PRACTICE PROBLEM 5.2

The transition from one to two equations can be particularly difficult for those who are new to numerical integration. Using the midpoint method, numerically integrate the following two equations:

$$\frac{dm}{dt} = p$$

$$\frac{dp}{dt} = 2m$$

given that $m(0) = 4$, $p(0) = 2$, $t(0) = 0$, and $\Delta t = 1$.

SOLUTION

First, find m_m and p_m. The equations are:

$$m_m = m_0 + f_m(m_0, p_0) \cdot \frac{\Delta t}{2}$$

$$p_m = p_0 + f_p(m_0, p_0) \cdot \frac{\Delta t}{2}$$

where $f_m = p$, and $f_p = 2m$. Substitution yields:

$$m_m = m_0 + (p_0) \cdot \frac{\Delta t}{2}$$

$$m_m = 4 + 2 \cdot \frac{1}{2} = 5$$

$$p_m = p_0 + (2m_0) \cdot \frac{\Delta t}{2}$$

$$p_m = 2 + (2 \cdot 4) \cdot \frac{1}{2} = 6$$

Next, use m_m and p_m to find m_1 and p_1:

$$m_{1,m} = m_0 + f_m(m_m, p_m) \cdot \Delta t$$

$$m_{1,m} = 4 + (6) = 10$$

$$p_{1,m} = p_0 + f_p(m_m, p_m) \cdot \Delta t$$

$$p_{1,m} = 2 + (2 \cdot 5) = 12$$

THE RUNGE–KUTTA METHOD

We can carry the principles of the midpoint and Euler methods still further using a higher-order approximation. Carl Runge and Martin Kutta, of the Technische Hochschule Hannover and the Rheinisch-Westfälische Technische Hochschule Aachen, respectively, developed a robust implementation of this approach, which is named for them. As we have seen, the error in both the Euler and midpoint methods essentially depends on an imperfect approximation of the slope (Figure 5.6). The Runge–Kutta method addresses this problem using a weighted average of slopes.

The first slope, which we will call f_1, is the slope calculated using the Euler method:

$$f_1 = f(x_0) \tag{5.18}$$

In other words, f_1 is simply the slope of the curve at x_0 (Figure 5.8, left). To simplify the equations, let's define a set of variables k as $k_i = \Delta t \cdot f_i$. The second slope is then calculated using the following equation:

$$f_2 = f\left(x_0 + \tfrac{1}{2}k_1\right) \tag{5.19}$$

If you compare Equation 5.19 with Equation 5.17, you will see that the second slope is the slope that would be calculated using the midpoint

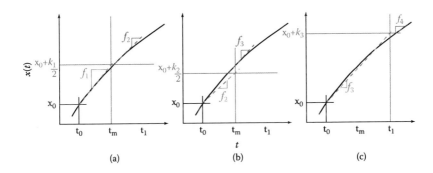

FIGURE 5.8 Calculation of the four slopes for implementation of the Runge–Kutta method. Left, f_1 is the slope of the curve at x_0; f_2 is the slope that would have been calculated using the midpoint method in Figure 5.6. Middle, f_3 is also calculated via the midpoint method, but with f_2 in place of f_1. Right, f_4 is based on the estimated end point. These four calculated slopes are weighted and averaged to estimate x_1.

method (Figure 5.8, left). The third slope is also calculated via the midpoint method (Figure 5.8, middle), with the exception that the second slope is used in the calculation in place of the first:

$$f_3 = f\left(x_0 + \tfrac{1}{2}k_2\right) \tag{5.20}$$

Finally, the fourth slope is not determined at the beginning of the time interval, or at estimated midpoints, but at an estimated end point (Figure 5.8, right):

$$f_4 = f(x_0 + k_3) \tag{5.21}$$

The four slopes are weighted and averaged to calculate the new value of x_1:

$$x_{1,RK} = x_0 + \tfrac{1}{6}(k_1 + 2k_2 + 2k_3 + k_4) \tag{5.22}$$

Figure 5.9 depicts the analytical solution for Equation 5.5, together with the three numerical methods discussed in this chapter. In this figure, the time step was equal to 1, and so only six points were calculated for each numerical method (your comparison of the methods in Figure 5.9 should focus on the points, not the lines between the points). You should see that for these six points, the Runge–Kutta solution

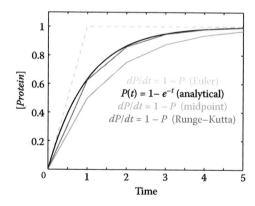

FIGURE 5.9 Comparison of the analytical solution (black) with the Euler (dashed light gray), midpoint (gray), and Runge–Kutta methods (dark gray) for $\Delta t = 1$. The Runge–Kutta method generates a curve that is extremely similar to the analytical solution.

TABLE 5.2   Values of k Calculated
Using the Runge–Kutta Method

i	$k_i = \Delta t \cdot f(x_{i-1})$
1	$1 - 0 = 1$
2	$1 - 1 \cdot \frac{1}{2} = \frac{1}{2}$
3	$1 - \frac{1}{2} \cdot \frac{1}{2} = \frac{3}{4}$
4	$1 - \frac{3}{4} = \frac{1}{4}$

is nearly indistinguishable from the analytical solution; the other numerical methods are clearly different.

To help illustrate the Runge–Kutta method more clearly, let's calculate the value of [Protein] at $t = 1$, [Protein]$_0 = 0$, and $f([Protein]) = 1 - [Protein]$, as given in Equation 5.5. With $\Delta t = 1$, we can simply consider the four slopes in Table 5.2 using Equations 5.18–5.21.

With these values, x_1 is determined using Equation 5.22 as $0 + 1/6 \cdot (1 + 1 + 6/4 + 1/4) = 5/8$ or 0.625. For comparison, the value of $x_{t=1}$ determined analytically is ~0.632. Using Equation 5.8 to estimate the error of the Runge–Kutta method for this particular case yields a value of 0.0175, which is 33-fold more accurate than the Euler method and over 20-fold more accurate than the midpoint method. Furthermore, the Taylor series approximation approach to determining error reveals that the Runge–Kutta method error is proportional to Δt^5.

The Runge–Kutta method is therefore fourth order and in fact, is the 4 in MATLAB's ode45 function. The 5 comes from adaptively controlling the step size, which is a nice addition but often unnecessary. For most uses, the standard Runge–Kutta method has stood the test of time.

You now have all three major tools for modeling and understanding ODEs in your tool belt—congratulations! These numerical approaches are particularly important because the size of the system you can study increases dramatically. But why did I force you to learn something that MATLAB can easily do for you? The answer is that knowing how the algorithm works will give you the power to change it when necessary. Without this ability, my lab would never have been able to create a model of a whole cell, for example, because we wanted to integrate ODE-based approaches with other approaches as far-flung as Boolean logic, linear optimization, and stochastic methods (Chapters 9 and 10). This kind of integration is discussed further, and you will see that your familiarity with these numerical methods will be critical.

CHAPTER SUMMARY

Numerical integration is the most common way to solve sets of ODEs, in particular when more than a few ODEs need to be solved simultaneously. We first considered the Euler method; we began with a set of initial conditions and then used the instantaneous rate of change at those conditions to estimate the new conditions. For a one-dimensional ODE, we showed how the instantaneous rate of change gives us the slope of a line, which we multiplied by the time step length and added to the initial value (at t_0) to obtain the next value (at t_1). The Euler method is highly sensitive to the step size (the error is supposed to be proportional to the square of the time step, as we determined by comparison to a Taylor series expansion), so other methods have been devised that are associated with a smaller error. The Euler method is called a first-order method because the error is proportional to the square of the time step.

The midpoint method uses the Euler method to estimate the variable value at t_m (halfway between t_0 and t_1) and to determine the instantaneous rate of change at that value to generate a slope. The midpoint method is called second order because its error is estimated to be proportional to the cube of the time step.

The Runge–Kutta method builds on the Euler and midpoint methods by calculating a weighted average of four slopes (the Euler and midpoint slopes, together with a second estimate of the instantaneous rate of change at t_m based on the midpoint, and an estimate of the instantaneous rate of change at t_1 based on the third slope). The error associated with the Runge–Kutta method is proportional to the fifth power of the time step, so this method is called fourth order. This method is a commonly used numerical integrator and has been implemented in MATLAB (with a variable step size) as `ode45`. Even though this function is simple to implement, it is important for us to learn how numerical integration works so that we can adapt the methods to our own purposes as necessary.

RECOMMENDED READING

Ellner, S. and Guckenheimer, J. *Dynamic Models in Biology*. Princeton, NJ: Princeton University Press, 2006.

Moore, H. *MATLAB for Engineers* (3rd edition). Englewood Cliffs, NJ: Prentice Hall, 2011.

Press, W. H., Teukolsky, S. A., Vetterling, W. T., and Flannery, B. P. *Numerical Recipes Third Edition: The Art of Scientific Computing*. Cambridge, UK: Cambridge University Press, 2007.

PROBLEMS

PROBLEM 5.1
A Simple ODE

Problem 5.1 is more illustrative than biological in order to build your skills. Consider the differential equation

$$\frac{dp}{dt} = p^3$$

with the initial condition $p_0 = 0.5$ at $t = 0$.

a. Calculate the value of $p(t = 1 \text{ s})$ by hand using the Euler, midpoint, and Runge–Kutta methods.

b. Now, solve the equation analytically. How does the analytical solution compare with your answers from (a)?

c. Write MATLAB code to simulate the dynamics of p over time using the Runge–Kutta method. Simulate 1 s with $\Delta t = 0.01$ s, beginning with $p_0 = 0.5$. Plot your result. Did the smaller time step make the Runge–Kutta calculation more accurate?

PROBLEM 5.2
Two Simple ODEs

Using the Runge–Kutta method, numerically integrate the following two equations:

$$\frac{dm}{dt} = p$$

$$\frac{dp}{dt} = 2m$$

given that $m(0) = 4$, $p(0) = 2$, $t(0) = 0$, and $\Delta t = 1$.

a. Calculate by hand the values of $m(t = 1)$ and $p(t = 1)$. Show your work.

b. Write MATLAB code to simulate the dynamics of m and p over time for 100 time steps with $\Delta t = 0.1$. Don't use ode45, of course! Plot your result.

c. At a party, a crowd of people watches from behind as you code your own numerical integrator. Under that pressure, you forget how to code a Runge–Kutta integrator and are forced to use the Euler method. What parameter would you change to make your integrator more accurate? What negative consequences would that have compared to higher-order methods?

PROBLEM 5.3
Our Favorite Autoregulatory, Negative-Feedback System

Recall that the system we have been studying throughout these chapters is governed by two ODEs (Equations 4.7 and 4.8):

$$\frac{d[Protein]}{dt} = k_{trl}[mRNA] - k_{ploss}[Protein]$$

$$\frac{d[mRNA]}{dt} = k_{trs,max}\left(\frac{1}{1+[Protein]/K}\right) - k_{mloss}[mRNA]$$

Let $K = k_{ploss} = k_{trl} = k_{trs,max} = k_{mloss} = 1$ and start with $[Protein]_0 = [mRNA]_0 = 0$.

a. Write three numerical integrators to solve this pair of ODEs using the Euler, midpoint, and Runge–Kutta methods. Use $\Delta t = 0.01$ and integrate from $t = 0$ to $t = 10$. Plot $[mRNA]$ and $[Protein]$ on the same plot for each integrator.

b. Now, run the same simulations at larger time steps, $\Delta t = 0.1, 0.5, 1.0,$ and 5.0. How do your integrations compare with one another at these different time steps?

PROBLEM 5.4
A Single Feedback Loop System with Enzymatic Degradation

Problem 4.3 introduced you to the use of dynamic protein degradation to create a bistable switch using the Lon protease found in *Mesoplasma florum* (*mf*-Lon). As mentioned in Chapter 4, the positive autoregulatory loop is identical to the one found in lambda phage, for which promoter P_{RM} expresses the lambda repressor cI, which is also a transcriptional activator of P_{RM}.

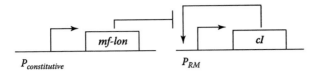

The equations that govern the system depicted in the figure are:

$$\frac{d[m_{cl}]}{dt} = k_{trs,\,max}\left(\frac{1}{1+\left(\dfrac{K_{Diss}}{[p_{cl}]}\right)^{H}}+l_{PRM}\right)-k_{mloss}[m_{cl}]$$

$$\frac{d[p_{cl}]}{dt} = k_{trl}[m_{cl}]-k_{ploss}[p_{cl}]-k_{cat}[p_{L}]\left(\frac{1}{1+\dfrac{K_{M}}{[p_{cl}]}}\right)$$

where H equals the Hill coefficient of the system, K_{Diss} is the dissociation constant of cI from P_{RM}, l_{PRM} is the fraction of mRNA transcription from the P_{RM} promoter that is caused by leakage (basal expression), k_{cat} is the catalytic rate of mf-Lon, p_{L} is the protein concentration of mf-Lon, and K_{M} is the Michaelis constant for mf-Lon (Michaelis constants are discussed in more detail in Chapter 9). As described in Problem 3.4, when multiple repressor binding sites occur in front of a gene, the Hill coefficient describes the cooperative binding properties of that system (for two binding sites, $H = 2$, etc.).

As before, let's use the following parameter values:

$$k_{trs,\,max} = 1.35 \times 10^{-9}\ \text{M}^{2}\ \text{s}^{-1}$$

$$K_{diss} = 2.5 \times 10^{-8}\ \text{M}$$

$$l_{PRM} = 0.1$$

$$k_{mloss} = 2.38 \times 10^{-3}\ \text{s}^{-1}$$

$$k_{trl} = 5 \times 10^{-5}\ \text{s}^{-1}$$

$$k_{ploss} = 2 \times 10^{-4}\ \text{s}^{-1}$$

$$k_{cat} = 0.071 \text{ s}^{-1}$$

$$p_L = 1 \times 10^{-10} \text{ M}$$

$$K_M = 3.7 \times 10^{-9} \text{ M}$$

$$H = 2$$

Let the initial amounts of cI mRNA be 200 nM and cI protein be 50 nM.

a. Calculate the steady-state quantities of cI mRNA and protein analytically.

b. Using the midpoint method, find the concentrations of mRNA and protein at $t = 0.1$, using $\Delta t = 0.1$. Show your work.

c. Now build a Runge–Kutta numerical integrator in MATLAB to solve the ODEs. Generate plots of mRNA and protein concentrations over time. How do your results compare to your answer for (a)?

d. Plot the nullclines for the two ODEs (hint: you may have already done this as part of Problem 4.3). Now, on the same nullcline graph, plot the paths taken by m_{cI} and p_{cI}. Make sure you run enough time steps to visualize the behavior you expect! You will find the hold command to be useful here.

e. Change the initial conditions as follows, plotting the concentrations of cI mRNA and protein over time and also on the nullcline plot: Try $[m_{cI}] = 70$ nM and $[p_{cI}] = 10$ nM, and then $[m_{cI}] = 151.53875097$ nM and $[p_{cI}] = 11.2000778972$ nM (yes, I chose these specific numbers on purpose). Are the results what you expect? Again, you will find the subplot command useful here.

CHAPTER 6

Variation: Stochastic Simulation

LEARNING OBJECTIVES

- Explain how a system that only involves a small number of molecules can require a different modeling approach

- Understand the physical basis for the Gillespie algorithm

- Implement stochastic simulations for sets of chemical reactions

- Determine the relationship between rate parameters based on ODEs and those based on stochastic simulation

We're almost finished with our simple circuit, but there is one thing we have not considered in detail yet: the individual cell. For example, all of the ODE-based approaches described in Chapters 3–5 depended on assumptions of mass action kinetics. These assumptions generally hold in a well-mixed system, in which the concentrations of reactants are uniform throughout the system. The population-averaged behavior of a large culture of individual cells also seems to be well modeled using such assumptions.

SINGLE CELLS AND LOW MOLECULE NUMBERS

However, the individual cell is not a well-mixed system, especially at very low numbers—it is impossible to "mix" a single molecule so that it is equally distributed throughout a cell. This fact can lead to interesting biology, as elegantly demonstrated by Sunney Xie's lab at Harvard University, where scientists used cutting-edge technology to determine the protein count and the mRNA count for thousands of genes in thousands of individual *E. coli* cells. These data were published in *Science* in 2010 (Taniguchi et al., 2010).

Xie's lab's approach is worth describing because it was very clever. First, they built a library of *E. coli* strains in which individual genes had been fused to the yellow fluorescent protein gene (*yfp*, encoding YFP). They then calculated the number of expressed fluorescent proteins per cell using microscopy and image analysis. To count the number of mRNAs of the same gene in the same cell, the lab used a technique called fluorescence *in situ* hybridization, or FISH. In FISH, a number of short **oligonucleotides** (generally 30–50 nucleotides each) are constructed that are complementary to the mRNA of interest. Each "oligo" is bound to a fluorescent probe, so that when a given mRNA is present, the oligos hybridize to it (because of their complementary sequences) and can be detected with microscopy. One key insight here was to only develop oligos that hybridized to the *yfp* mRNA instead of to all of the other genes one by one, which saved a lot of money and time.

With this data set, the Xie lab was positioned to ask a fundamental question that no one had ever been able to address: What is the relationship between the number of mRNAs and the number of proteins in individual cells? Their answer was shocking at the time, and you can see it for yourself in Figure 6.1: *There is no evident relationship between the two.*

Why is this the case? After all, we already know that transcription and translation are related through the central dogma. This quandary is an interesting case of "causation without correlation," and the difference has to do with the stability of the mRNA molecule. As we discussed in Chapter 2, *E. coli's* mRNA degrades rapidly, with an average half-life of a few minutes. In contrast, each mRNA is assumed to direct the production of ~10 proteins, each of which remains stable for more than 10 h on average. This huge difference in the timescales of decay leads to a protein pool that is much larger than the associated mRNA pool, which you can see from the axes of the plot at the bottom of Figure 6.1; notice that there are thousands of proteins, while the number of mRNAs ranges from zero to ten. This small number of mRNA molecules that are in the cell at any given time adds a strong component of random variability to our observations because the same process, working essentially in the same way but in two different cells, can lead to two very different readouts.

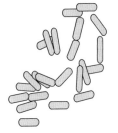
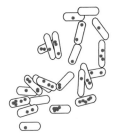

YFP protein excitation at 514 nm   Excitation of mRNA label at 580 nm

(a)

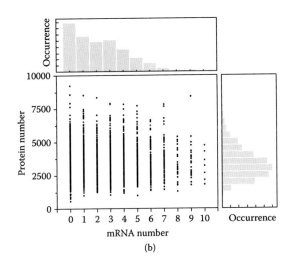

(b)

FIGURE 6.1 There is no relationship between protein number and mRNA number in the same cell at a given time. (a) The protein (left) and mRNA (right) molecule numbers were counted with fluorescent protein fusions and FISH, respectively. (b) Pattern emerges from the protein and mRNA counts for the *tufA* gene; each point is the count for a single cell (more than 5,000 cells are represented in this dataset). (From Taniguchi, Y., Choi, P. J., Li, G. W., Chen, H., Babu, M., Hearn, J., Emili, A., and Xie, X. S. *Science* 2010, **329**(5991): 533–538. Reprinted with permission from AAAS.)

STOCHASTIC SIMULATIONS

How can we model these observations? The modeling approaches that we have studied so far are **deterministic**, meaning that the same initial conditions input into the same model will lead to the same dynamic response and outputs. In contrast, to capture the natural variability we see in Figure 6.1, we need a method that is stochastic, meaning that it accounts for random events.

Stochastic methods differ from ODEs in a number of ways, some of the most important of which are listed in Figure 6.2. In general, ODEs are the best choice when the system contains many molecules. Stochastic methods become important when the number of molecules is very small because the concept of "concentration" breaks down (because there is only one molecule, we need to consider the odds that the molecule will be in the right place at the right time) and each molecule needs to be counted individually. Accordingly, the data produced by an ODE-based model are **continuous**; the data produced by a stochastic method are **discrete**. In addition, the concept of a reaction "rate," which we used for ODEs, must be substituted with the probability that a reaction occurs over a given time interval (we return to this idea in further discussion).

All of the conceptual differences between ODEs and stochastic methods lead to a substantial change in the way we frame our modeling problem, even for the same system. One way to frame the general problem of modeling the reactions and interactions of species (such as proteins, DNA, and RNA) over time as an ODE is as follows:

Given the concentrations of a number of species x_i at a given time point t_0, as well as the functions $dx_i/dt = f(x,t)$, determine x_i at a later time point $t_0 + \Delta t$.

Meanwhile, the stochastic problem is represented more like this:

Given a fixed volume V containing a number N of chemical species S_i, which are present with counts X_i at a given time point t_0, and given that these species can interact through a number M of chemical reactions R_j, simulate X_i at a later time point $t_0 + \Delta t$.

ODEs	Stochastic methods
Many molecules	Few molecules
Concentrations	Counts
Continuous	Discrete
Rates	Probabilities/time

FIGURE 6.2   A comparison of ODEs and stochastic methods.

The stochastic formulation contains a slew of variables and symbols, so the best way to compare these statements directly is with an example (Figure 6.3). Figure 6.3a shows two reactions, and Figure 6.3b gives the ODEs and initial concentrations that we might assemble given the reaction system. Here, the number of chemical species N is equal to 3, and the number of chemical reactions M is equal to 2. The table in Figure 6.3c contains both the identity of species S_i as well as the exact number X_i of each molecule at time t_0, and the table in Figure 6.3d represents the two reactions. Note that there are also constants c_j for each reaction, comparable but rarely equal to the ODE rate constants k_j. We return to this point later in the chapter.

Using this framework, in 1976 Daniel Gillespie developed a stochastic simulation approach that has been widely useful in many fields, including systems biology. The core of the idea is to answer two questions about the system at a given time point to determine what happens in the future. The first question is, *When does something happen next?* The second question naturally follows: *When that moment arrives, what actually happens?* In other words, given the number of molecules of each species, the volume of the space, and the available chemical reactions

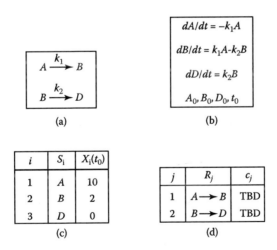

FIGURE 6.3   Setting up ODEs and a stochastic simulation for a simple reaction system. (a) Two reactions convert a single reactant (left) to a single product (right). (b) A set of ODEs and initial concentrations describing the system. (c) Table of the species in the system, including the number of molecules of each species at a given time (I chose the numbers 10, 2, and 0 simply as an example). (d) Table of reactions. The constants c are to be determined (TBD) later in this chapter.

and their associated likelihoods, which molecules react next and when? The answers to these two questions determine the trajectory of the stochastic simulation.

THE PROBABILITY THAT TWO MOLECULES INTERACT AND REACT IN A GIVEN TIME INTERVAL

Answering these two questions requires some mathematical gymnastics. Let's start by considering something simpler. Assume that we have one molecule each of two species, S_1 and S_2, bouncing around in a volume V (Figure 6.4a). Each molecule has a position within V, and the molecules have radii r_1 and r_2, respectively. Furthermore, we consider the two velocities of each by defining a vector v_{12}, which is the relative velocity of S_1 to S_2.

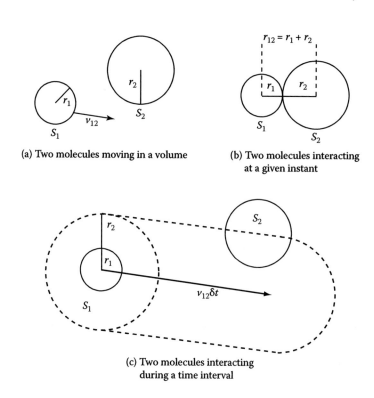

(a) Two molecules moving in a volume

(b) Two molecules interacting at a given instant

(c) Two molecules interacting during a time interval

FIGURE 6.4 Defining the interaction between two molecules. (a) One molecule each of species S_1 and S_2, with associated radii (r_1 and r_2, respectively) and the relative velocity of S_1 to S_2 (v_{12}). (b) An interaction occurs when r_{12}, the distance between S_1 and S_2, is equal to the sum of the individual radii. (c) A collision volume is defined by the two radii as well as the relative velocity and the time interval, δt.

We want to determine the probability that these two molecules will interact and react during a given time interval δt. To begin, we note that the interaction of S_1 and S_2 can be defined as occurring when the distance between the centers of each molecule is equal to the sum of the radii, $r_1 + r_2$ (when the molecules are touching; Figure 6.4b). As a result, we can define a "collision volume" δV_{coll} as a cylinder (Figure 6.4c):

$$\delta V_{coll} = \pi r_{12}^2 \left|\overrightarrow{v_{12}}\right| \delta t \tag{6.1}$$

If the center of the S_2 molecule lies within the S_1 molecule's collision volume during the time interval, then the molecules will collide and interact; otherwise, no collision will occur during that time.

Defining a collision volume makes it easy for us to calculate the probability that these two molecules will interact; we simply want to know how likely the S_2 molecule is to lie within δV_{coll}. If we assume that both molecules are equally likely to be in any part of the total volume, then the probability of an interaction between the S_1 and S_2 molecules during the given time interval is simply the ratio of the collision volume to the total volume:

$$\text{Prob}(S_1\text{-}S_2 \text{ interaction}) = \delta V_{coll}/V \tag{6.2}$$

Now let's apply this result more broadly: Given a mixture of multiple molecules of S_1 and S_2, what is the probability that a molecule of S_1 will react with a molecule of S_2? Here, we take advantage of two assumptions. First, we assume that not many molecules are in the system. The assumption sounds reasonable, since this is why we're pursuing stochastic simulations in the first place. The second assumption is that the time step is short. Taken together, these assumptions imply that, for a given δt, the probability of finding two or more molecules of S_2 in any S_1 collision volume is vanishingly small. In other words, the outcome for a given molecule of S_1 in a given time step is binary: either it interacts with a single molecule of S_2 or it does not.

So for any given S_1-S_2 pair in the mixture, we can use Equation 6.2 to determine the probability that they will interact. We can then calculate the average probability that any S_1-S_2 pair interacts from the average velocities:

$$\left\langle \frac{\delta V_{coll}}{V} \right\rangle = \frac{\pi r_{12}^2 \left\langle \overrightarrow{v_{12}} \right\rangle \delta t}{V} \tag{6.3}$$

where the pointed brackets denote averages. If we assume that the distribution of molecular velocities in the system follows a particular distribution (in Gillespie's case, he assumed that the velocities followed the **Boltzmann distribution**), then you could substitute an equation for the average based on that distribution, but we do not need to do that here.

Instead, let's consider how many of the S_1-S_2 pairs in the system that *interact* will also *react*. For example, with the Boltzmann distribution assumption, we would use the following equation:

$$\text{Prob}\left(\text{reaction} \,|\, \text{interaction}\right) = \exp\left(\frac{-U_u^*}{KT}\right) \qquad (6.4)$$

where U_u^* is the activation energy for the reaction, K is Boltzmann's constant, and T is the absolute temperature of the system. The right side of Equation 6.4 will always be in the range of zero to one because the pairs that react are a subset of the pairs that interact. Substituting Equation 6.3 into 6.4, we obtain:

$$\text{Prob}(\text{reaction}) = \frac{\pi r_{12}^2 \left\langle \overrightarrow{v_{12}} \right\rangle \exp\left(\frac{-U_u^*}{KT}\right)}{V} \delta t \qquad (6.5)$$

Nearly all of the terms in Equation 6.5 can be lumped together as a single constant c:

$$\text{Prob}(\text{reaction}) = c \, \delta t \qquad (6.6)$$

This constant, c, is the same constant shown in Figure 6.3d and expresses the probability that a given pair of molecules will react over a given time interval.

THE PROBABILITY OF A GIVEN MOLECULAR REACTION OCCURRING OVER TIME

Now that we know the probability of an S_1-S_2 pair of molecules reacting in a mixture, we can finally determine the total probability of the S_1-S_2 reaction. If we assume that the reaction is characterized by a constant c_1, then the probability of an average pair reacting is $c_1 \delta t$. To calculate the overall probability of the reaction, we only need to consider all of the possible pairs. For example, if there are X_1 molecules of S_1 and X_2 molecules of S_2, then the total number of possible unique pairs, sometimes denoted by an h, is $X_1 \cdot X_2$, and the total probability of this reaction occurring is:

$$\text{Total Prob(reaction}_1) = c_1 X_1 X_2 \delta t = c_1 h_1 \delta t = a_1 \delta t \qquad (6.7)$$

Notice that the subscripts for c, h, and a all refer to the reaction itself; the subscripts for the X variables refer to the molecular species. The variable a, sometimes referred to as the **reaction propensity**, is defined as the probability that the reaction will occur over a given time interval. The variable a is therefore analogous to the reaction rates that are used for ODEs. Also, notice that the form of the activity, $c_1 X_1 X_2$, is similar to the assumption of mass action kinetics in ODEs. In both cases, the rate or activity of the reaction is proportional to the product of the amounts of reactants.

THE RELATIONSHIP BETWEEN KINETIC AND STOCHASTIC CONSTANTS

These formulations may lead you to suspect that the c parameters in stochastic simulations are related to the k parameters in ODEs, and you would be right—but like most relationships, it's complicated! In his analysis, Gillespie (1976) defined the deterministic kinetic rate constant k as the average reaction rate per unit volume divided by the product of the average densities of reactants. In the case of many molecules and a well-mixed system, we would simply call the latter the concentrations of S_1 and S_2. Applying this definition to our simple system gives:

$$k_1 = \frac{\langle c_1 h_1 / V \rangle}{\langle X_1 / V \rangle \langle X_2 / V \rangle} = \frac{c_1 \langle X_1 X_2 \rangle / V}{\langle X_1 / V \rangle \langle X_2 / V \rangle} \qquad (6.8)$$

where the numerator comes from Equation 6.7. Rearranging Equation 6.8, we obtain:

$$k_1 = \frac{\langle X_1 X_2 \rangle}{\langle X_1 \rangle \langle X_2 \rangle} c_1 V \qquad (6.9)$$

Again, in a dense and well-mixed system, $<X_1 X_2>$ is equal to $<X_1><X_2>$, so:

$$k_1 = c_1 V \qquad (6.10)$$

In other words, the kinetic rate constant in this case is related to the stochastic simulation constant by the volume. This should make intuitive sense because kinetics are based on concentrations, but the simulations described here are based on the numbers of molecules.

I want to emphasize that Equation 6.10 is not general; it applies to this case but certainly not to others. For example, let's consider a second reaction, in which two molecules of S_1 react to form S_2:

$$S_1 + S_1 \xrightarrow{c_2} S_2 \qquad (6.11)$$

As you can see, this reaction is characterized by its own reaction constant, c_2. How does c_2 relate to its corresponding kinetic constant k_2? The major difference lies with how h_2 is calculated. Remember that h contains all of the possible reactions between molecules, in this case pairs of S_1. Since the two reactants are the same species, h is calculated differently:

$$h_2 = \binom{X_1}{2} = \frac{X_1(X_1 - 1)}{2!} \qquad (6.12)$$

The corresponding k for this reaction would therefore be defined as:

$$k_1 = \frac{\langle c_2 h_2 / V \rangle}{\frac{\langle X_1 \rangle}{V} \frac{\langle X_1 \rangle}{V}} = \frac{c_2 \langle X_1(X_1 - 1)/2 \rangle / V}{\frac{\langle X_1 \rangle}{V} \frac{\langle X_1 \rangle}{V}} = \frac{1}{2} c_2 \frac{\langle X_1 - 1 \rangle}{X_1} V \approx \frac{1}{2} c_2 V \qquad (6.13)$$

As shown, a simple change in the reaction structure leads to a twofold change in the relationship between k and c. The changes can also become more complicated; adding more reactants can lead to multiple powers of V in the equation, for example.

GILLESPIE'S STOCHASTIC SIMULATION ALGORITHM

We are finally ready to address the two questions posed at the beginning of the chapter. First, take another look at Equation 6.7, in particular the reaction activity a. This quantity is the probability that a given reaction occurs over a given time interval. Therefore, the reciprocal $1/a_j$ is the mean time required for the reaction R_j to occur. We can then define a_{total} as the propensity for any of the possible reactions in the system to occur:

$$a_{total} = \sum_{j=1}^{M} a_j \qquad (6.14)$$

The reciprocal $1/a_{total}$ then becomes the mean time required for any of the reactions to occur.

Therefore, to answer Question 1—When is the next time that something will happen in our system?—we simply take the following three steps:

1. Let τ (tau) be the time interval between the current time and the next reaction.

2. Generate a random number $rand_1$ between 0 and 1 from a **uniform distribution**.

3. Determine τ from the following equation:

$$\tau = \frac{1}{a_{total}} \ln\left(\frac{1}{rand_1}\right) \tag{6.15}$$

The logarithmic term in Equation 6.15 is simply the conversion of a random number from a uniform distribution into a random number drawn from an **exponential distribution**, which favors shorter times over longer times. Multiplying this random number by the mean time gives us the answer to Question 1.

Question 2—Which reaction occurs at the new time?—is also determined from the a's and a second random number:

1. Generate a random number $rand_2$ between 0 and 1 from a uniform distribution.

2. Find the smallest q such that

$$\sum_{j=1}^{q} a_j > a_{total} \cdot rand_2 \tag{6.16}$$

Equation 6.16 requires further explanation, and Figure 6.5 helps. Think of a number line that has a length of a_{total} and divide the line into its components a_j as seen in Equation 6.14. Now imagine that this number line is a one-dimensional dartboard. You throw a dart by multiplying a_{total} by $rand_2$. The result will fall somewhere on the a_{total} number line—in particular, in one of the a_j regions. The region that is selected corresponds to the reaction that occurs.

The way that we answer these questions, particularly by choosing random numbers, should reflect the fact that in this chapter we're

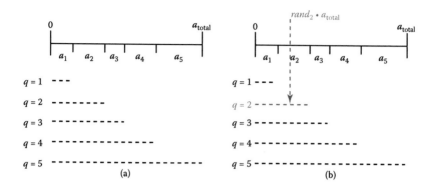

FIGURE 6.5 Understanding Equation 6.16. (a) Depicting a_{total} as a number line, as well as the left side of Equation 6.16 for all possible values of q given a five-reaction system for which a_j is known. (b) Every random number between 0 and 1 corresponds to a specific value of q.

performing stochastic simulations, which are different from the deterministic approaches that we covered previously. No two stochastic simulations will be exactly alike.

In pseudocode, the algorithm looks something like this:

```
Initialize time t₀, molecule numbers Xᵢ
Define cⱼ's for each reaction
For each time step:          %not uniform length!
        Compute aⱼ (c₁X₁X₂)
        a_total = sum(aⱼ)
        Choose two random numbers
        Find tau                %Equation 6.15
        Find q                  %Equation 6.16
        t = t₀ + tau
        Update Xᵢ's
end
end
```

Remember that the time steps are not uniform in length; their length is determined as τ.

Let's apply this methodology to an example with only two simple reactions:

$$A \xrightarrow{c_1} B$$

$$B \xrightarrow{c_2} D$$

First, we initialize the system. In this case, let $t_0 = 0$, $X_A(t_0) = 10$ molecules, $X_B(t_0) = 2$ molecules, and $X_D(t_0) = 0$ molecules. Now assume that the constant associated with the first reaction, c_1, will be equal to the second, $c_1 = c_2 = 0.5$.

To work through the next time step, we need to compute a_1 and a_2. From the reactions, we see that:

$$a_1 = c_1 X_A = 0.5 \cdot 10 = 5$$

$$a_2 = c_2 X_B = 0.5 \cdot 2 = 1$$

We can then calculate $a_{total} = a_1 + a_2 = 5 + 1 = 6$, for the six sections of our system. Now let's address the random numbers. Confession: The two numbers that I am about to generate are not "random" by any stretch of the imagination, but they are helpful for the example:

$$rand_1 = 1/e$$

$$rand_2 = 0.4$$

Using $rand_1$ and a_{total}, τ is calculated from Equation 6.15 as $1/6 \cdot \ln(e) = 1/6 = 0.167$. We can then find q using Equation 6.16, where $a_{total} \cdot rand_2 = 2.4$. Thus, the smallest q that exceeds 2.4 is $q = 1$, because $a_{q=1} = 5$ (Figure 6.6).

In other words, the first reaction, $A \rightarrow B$, is the one that occurs. This reaction consumes one A molecule and creates one B molecule. We therefore need to update our X_A and X_B values as follows:

$$X_A(t = \tau) = X_A(t_0) - 1 = 9$$

$$X_B(t = \tau) = X_B(t_0) + 1 = 3$$

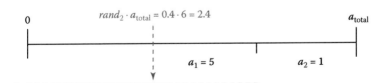

FIGURE 6.6 Updating Figure 6.5 for our example with only two simple reactions.

If the second reaction had been chosen instead, then we would decrement the number of Bs and increment the number of Ds. One time step of the stochastic simulation has been completed.

STOCHASTIC SIMULATION OF UNREGULATED GENE EXPRESSION

Let's apply this method to our favorite circuit! We start with the unregulated system, in which mRNA is produced by transcribing a gene, and this mRNA is translated to form protein. We can represent these processes as chemical reactions:

$$DNA \xrightarrow{c_{trs}} mRNA + DNA$$

$$mRNA \xrightarrow{c_{trl}} mRNA + Protein$$

Note that neither the DNA nor the mRNA is destroyed by the reaction, so they are both reactants and products. We now add reactions for the loss of mRNA and protein by decay. In this case, the right side of the reaction is empty to represent the loss of a molecule:

$$mRNA \xrightarrow{c_{mloss}}$$

$$Protein \xrightarrow{c_{ploss}}$$

Now I'm going to take you through actual MATLAB code to simulate the behavior of this system. Let's start with the header information and assignment of the c constant values. We assume that c_{trs} has a value of 0.005 mRNAs produced per second, which corresponds to 1 mRNA produced every few minutes or 18 molecules produced/h. For c_{trl}, a value of 0.167 proteins/mRNA/s yields 10 proteins/min/mRNA. Recall that the half-lives of mRNA and protein are quite different, so we choose half-lives of 2 min for mRNA and 60 min for protein. The c_{mloss} and c_{ploss} values are therefore ln(2) mRNAs/120 s and ln(2) proteins/3,600 s, respectively.

```
function output = stochasticSim()
c_tsc = 0.005;
c_tsl = 0.167;
c_mloss = log(2)/120;
c_ploss = log(2)/3600;
```

The next section of code initializes the system and allocates storage for our outputs:

```
%initial values
t = 0;                    %time
mRNA = 0;
Protein = 0;
DNA = 1;

%storage
xx = zeros(2,50000);
tt = zeros(1,50000);
a = zeros(1,4);
```

Notice that only the value for DNA is nonzero. This value will remain constant throughout the simulation because we are simulating a single cell, with only a single copy of our gene of interest. The tt array holds all of the time information, the xx array holds the values of mRNA and Protein at the times given in tt, and the a array holds the value of a_j at every time step.

Let's start the simulation as a while loop. We introduce a counter to tell us when we've gone through 50,000 steps. It is generally not advisable to make the conditional part of the loop depend on clock time because the steps vary in size and can become very small as the number of molecules increases. Furthermore, I often downsample the data in practice; I only record one time point for every 10 that I generate to reduce the file size of my plots. After starting the while loop, we follow the pseudocode that I wrote previously to calculate the propensity values:

```
counter = 1;
while counter < = 50000
        %update propensities
        a(1) = c_tsc * DNA;          %transcription
        a(2) = c_tsl * mRNA;         %translation
        a(3) = c_mloss * mRNA;       %mRNA loss
        a(4) = c_ploss * Protein;    %protein loss
        a_total = sum(sum(a));
```

Calculating the value for τ is fairly straightforward, given what you have already read:

```
        r1 = rand;
        tau = (1/a_total) * log(1/r1);
        t = t + tau;
```

Next, we need to find q:

```
r2 = rand;
comparison = a_total * r2;
sum_as = 0;
q = 0;
for i = 1:4                    %over all four
                                 reactions
     sum_as = sum_as + a(i);   %Equation 6.16
     if sum_as > comparison
       q = i; break;
     end;
end;
```

Now we need to calculate the new values for mRNA and protein based on which reaction occurred:

```
if q == 1    %transcription
     mRNA = mRNA + 1;
elseif q == 2      %translation
     Protein = Protein + 1;
elseif q == 3      %mRNA decay
     mRNA = mRNA - 1;
elseif q == 4      %protein decay
     Protein = Protein - 1;
end
```

Finally, we store the values and finish the program:

```
x = [mRNA; Protein];
tt(1, counter) = t;
xx(:, counter) = x;
counter = counter + 1;
end;
```

An example output is shown in Figure 6.7. As you can see, although the circuit's behavior is nothing like what we observed using ODEs (Figure 6.1b), there are some similarities. For example, the steady-state protein level that we would calculate using ODEs would be:

$$[Protein]_{ss} = \frac{k_{trl}}{k_{ploss}}[mRNA]_{ss} = \frac{k_{trl}}{k_{ploss}}\frac{k_{trs,max}}{k_{mloss}} \qquad (6.17)$$

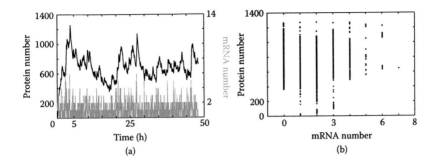

FIGURE 6.7 The mRNA and protein levels for our circuit, calculated using a stochastic simulation. (a) mRNA (gray) and protein (black) counts for a single simulation. Since these simulations are stochastic, this output may vary from simulation to simulation. (b) Instantaneous mRNA number plotted against the protein number at the same time point for the entire simulation (50,000 time points).

In our case, all of the k parameters are related to the c parameters in the same way because there is only a single reactant. For example, for the reaction DNA → mRNA + DNA:

$$k_{trs} = \frac{\langle c_{trs} DNA/V \rangle}{\langle DNA \rangle/V} = \frac{c_{trs} \langle DNA \rangle/V}{\langle DNA \rangle/V} = c_{trs} \qquad (6.18)$$

Therefore, we can substitute our k's for c's and vice versa, leading to an expected level of protein at steady state in the ODEs:

$$[Protein]_{ss} = \frac{0.167 \dfrac{[Protein]}{[mRNA] \cdot s}}{\dfrac{\ln(2)}{120\,s}} \cdot \frac{0.005 \dfrac{[mRNA]}{s}}{\dfrac{\ln(2)}{3600\,s}} = 750 \text{ proteins} \quad (6.19)$$

Looking at Figure 6.7, although the cell does not seem to reach a true steady state, the protein level appears to remain mostly between 600 and 1,000 molecules over time.

Let's return to the dynamics of the mRNA and protein. We're used to seeing dynamics that look like exponentials from the ODEs. In contrast, here the mRNA production and loss dynamics are highly stochastic, and this stochasticity has a major effect on the production and loss of protein. These protein production dynamics are sometimes called "burst-like" because the production of a single mRNA in the system often leads to a burst of subsequent translation and protein production. Look at the earliest part of the simulation; when mRNA shoots up to 4 molecules/cell, protein production also rises dramatically, but when the mRNA level goes down to 1 or 0 molecules/cell shortly thereafter, protein production also halts.

We can also revisit the Xie lab's observation (Figure 6.1b) that the mRNA and protein counts in any given cell at an instant in time do not seem to be correlated with one another. We can create an analogous plot by running the stochastic simulation many times. If we record the mRNA and protein counts from a randomly chosen time point in each simulation (Figure 6.7b), our plot looks strikingly similar to the experimental data.

PRACTICE PROBLEM 6.1

The previous example does not include negative feedback. Write the additional chemical reaction(s) to describe the interaction of the protein with the DNA, remembering that the protein-DNA complex does not support transcription. How would you change the pseudocode to reflect this change?

SOLUTION

The inclusion of negative feedback requires two additional reactions, one that leads to association of protein to the DNA to form a complex and another to reflect dissociation of the complex into free protein and DNA:

$$DNA + Protein \xrightarrow{\;c_{assoc}\;} Complex$$

$$Complex \xrightarrow{\;c_{dissoc}\;} DNA + Protein$$

Only free DNA supports transcription, so there is no need to add another reaction; we already have a transcription reaction involving the free DNA.

To implement these changes into our pseudocode, we first create storage for the DNA and complex by replacing the line for xx:

```
xx = zeros(4,50000);
```

Next, we need to estimate c_{assoc} and c_{dissoc}. For our example, let's use:

```
c_assoc = 0.00001; %bindings/DNA/protein/s
c_dissoc = 0.005; %unbindings/complex/s
```

If there are 1,000 proteins in a cell and one copy of the DNA binding site, the mean time for complexation to occur is (1 DNA · 1,000 proteins · c_{assoc})$^{-1}$ or 100 s. With only one complex in the cell, the mean time for dissociation to occur is (1 complex · c_{dissoc})$^{-1}$ or 200 s, twice as long. These numbers are convenient for us because I want you to visualize the complex from the plots, but in "real life" I would expect the binding and unbinding to occur much faster, probably by two to three orders of magnitude.

Note also that c_{assoc} is related to k_{assoc} by a factor of the volume, as you can convince yourselves using the equations already given. To make this conversion, we estimate that the typical E. coli cell is 1 μm^3 in volume (see Recommended Reading), so one molecule per cell corresponds to a concentration of roughly 1 nM. As a result, c_{assoc} = 0.00001 binding events/DNA molecule/protein molecule/s, and k_{assoc} = 0.00001/nM/s.

We update our propensities a_j with two new statements and a replacement for a_{total}:

```
a_assoc = c_assoc * Protein * DNA;
a_dissoc = c_dissoc * Complex;
a_total = a_tsc + a_tsl + a_degM + a_degP...
              + a_assoc + a_dissoc;
```

Finally, we add the species updates for association and dissociation:

```
elseif q == 5
    %complex formation
    Protein = Protein - 1;
    DNA = DNA - 1;
    Complex = Complex + 1;
elseif q == 6
    %complex dissociation
    Protein = Protein + 1;
    DNA = DNA + 1;
    Complex = Complex - 1;
end
```

Making these changes leads to the simulation outputs in Figure 6.8. Notice in the upper panel that mRNA is not made when the complex is formed, and as a result, the protein tends to decay.

In addition, we can run many simulations (100 in Figure 6.9) to see what happens to the protein and mRNA levels in aggregate, both with and without feedback (Figure 6.9). This figure has been plotted starting at $x = 0.5$ h for clarity. Notice that just as with the ODEs, the system with feedback has a lower steady-state protein count and a faster response time than the system without feedback.

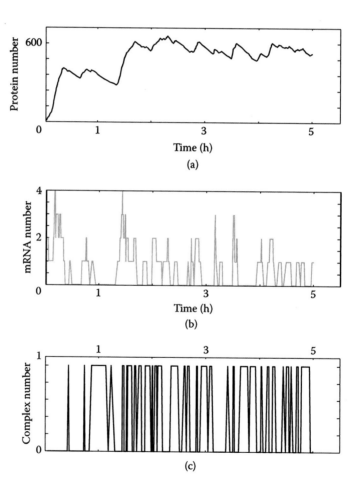

FIGURE 6.8 Adding feedback to the stochastic simulation in Practice Problem 6.1. Shown are the simulation outputs for the amounts of protein (a), mRNA (b), and DNA-protein complex (c) in a single simulation.

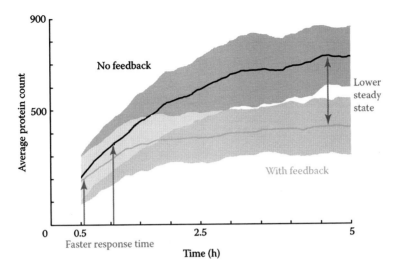

FIGURE 6.9   The average (lines) and distribution (shading) of protein count for 100 simulations. The circuit with no feedback appears in darker gray (top line), and the circuit with feedback appears in lighter gray (bottom line). Note that the system with feedback has both a faster response time (left) and a lower steady state (right). The first time point shown is at 0.5 h because the simulations are not fit well with a spline or linear interpolation at earlier time points.

STOCHASTIC SIMULATIONS VERSUS OTHER MODELING APPROACHES

As you can see, stochastic simulations are an extremely useful representation of our genetic system in the context of a single cell. These simulations not only helped us to better understand single-cell measurements, but also recaptured much of the information in the ODE-based models. So why not use stochastic simulations all the time? The more molecules and reactions in the system, the shorter the calculated time interval τ becomes; therefore, the simulation can quickly become too complex to compute. We can address this problem by creating hybrid models, in which some parts of the model are deterministic and use, for example, ODEs, while other parts that involve a small number of reacting molecules are simulated using a stochastic approach. We explore such hybrid models in later chapters.

In the meantime, you now have the essential systems biologist's tool kit—congratulations and thanks for sticking it out this far! Let's apply these methods to some relatively simple case studies and then move on to more complicated cellular processes.

CHAPTER SUMMARY

To this point, all of the methods considered work best for large concentrations of molecules, such as you would find when growing a large number of cells in a flask. However, modern technology has enabled us to make measurements within individual cells, where the number of molecules is much smaller. The behavior of a system with small numbers of molecules is characterized by random events, chance collisions between two molecular species that can interact and react with some probability, for example. These random occurrences are better modeled with stochastic simulations, and the classical method uses the Gillespie algorithm.

At any given instant in time, the Gillespie algorithm seeks to answer two questions: (1) When is the next reaction going to occur? and (2) Which of the possible reactions actually occurs at that time? To answer both questions, we first calculate the probability of occurrence over time for each reaction. Next, we sum the total probability for all reactions over time and calculate the mean time required for a reaction to occur as the reciprocal of the total probability over time. With this mean, we can define a distribution of times (based on an exponential distribution) and, by choosing a random number, select the time at which the reaction occurs. We then choose another random number from a uniform distribution and use this number to select which reaction occurs, with the likelihood of a reaction being selected proportional to its probability over time.

Stochastic simulations can be computationally expensive, especially as the number of molecules and reactions increases. As a result, their use is typically limited to smaller systems. Nevertheless, if they can be applied, these simulations can produce a more realistic estimation of single-cell behavior.

RECOMMENDED READING

BioNumbers. Home page. http://www.bionumbers.hms.harvard.edu. This website is a database of biological numbers, many of them approximations that are useful for the types of calculations performed in this book.

Gillespie, D. T. A general method for numerically simulating the stochastic time evolution of coupled chemical reactions. *Journal of Computational Physics* 1976, **22**(4): 403–434.

Gillespie, D. T. Exact stochastic simulation of coupled chemical reactions. *Journal of Physical Chemistry* 1977, **81**(25): 2340–2361.

Press, W. H., Teukolsky, S. A., Vetterling, W. T., and Flannery, B. P. *Numerical Recipes Third Edition: The Art of Scientific Computing*. Cambridge, UK: Cambridge University Press, 2007.

Taniguchi, Y., Choi, P. J., Li, G. W., Chen, H., Babu, M., Hearn, J., Emili, A., and Xie, X. S. Quantifying *E. coli* proteome and transcriptome with single-molecule sensitivity in single cells. *Science* 2010, **329**(5991): 533–538.

PROBLEMS

PROBLEM 6.1
Probabilities and Rate Constants

Previously, we calculated the probability of the reaction

$$S_1 + S_2 \rightarrow S_3$$

to be:

$$\text{Total Prob(reaction)} = c_1 X_1 X_2 \delta t$$

where X_1 and X_2 are the numbers of molecules of S_1 and S_2, respectively.

a. What are the probabilities of the following reactions?

Reaction 1: S_1 is degraded:

$$S_1 \rightarrow$$

Reaction 2: Two molecules of S_1 and one molecule of S_2 react to form S_3:

$$S_1 + S_1 + S_2 \rightarrow S_3$$

Reaction 3: Three molecules of S_1 react to form S_2:

$$S_1 + S_1 + S_1 \rightarrow S_2$$

b. For Reaction 3, also derive the relationship between the rate constant k and the stochastic constant c.

c. How does the Gillespie algorithm account for the possibility of a single molecule participating in two different reactions in the same time interval?

PROBLEM 6.2
Enzymatic Conversion

Consider a system in which a molecule X is transported into a cell, and where an enzyme E is expressed constitutively and can react with a pair of X molecules to form a Y molecule, as shown below:

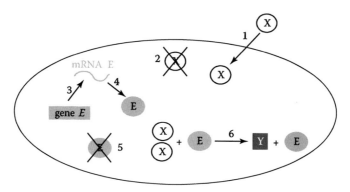

The system has six chemical reactions:

1. Import of X into the cell

2. Degradation of X inside the cell

3. Transcription, which produces an mRNA product

4. Translation, which requires an mRNA and produces an E protein

5. Degradation of E

6. The reaction of two internal X molecules in the presence of E to form Y

The stochastic rate constants are:

$$c_1 = 2/(\text{number of molecules of X}_{\text{outside}} \cdot \text{s})$$

$$c_2 = 3/(\text{number of molecules of X}_{\text{inside}} \cdot \text{s})$$

$$c_3 = 1/(\text{number of copies of gene } E \cdot \text{s})$$

$$c_4 = 1/(\text{number of mRNA molecules} \cdot \text{s})$$

$$c_5 = 1/(\text{number of molecules of E} \cdot \text{s})$$

$$c_6 = 2/(\text{number of molecules of E} \cdot \text{number of molecules of X}_{\text{inside}}^2 \cdot \text{s})$$

The initial conditions are:

$$X_{outside} = 10 \text{ molecules}$$

$$X_{inside} = 5 \text{ molecules}$$

$$E = 4 \text{ molecules}$$

$$\text{gene } E = 1 \text{ copy}$$

$$mRNA = 1 \text{ molecule}$$

a. Calculate a_1, a_2, a_3, a_4, a_5, a_6, and a_{total}.

b. What is the average time required for a reaction to take place?

c. What is the probability that X will be degraded under this set of conditions?

PROBLEM 6.3
Positive Autoregulatory Feedback Circuit with Cooperativity

In Problem 4.2, we constructed a set of ODEs for our simple autoregulatory circuit, but with positive instead of negative feedback and with cooperativity. The diagram is shown below:

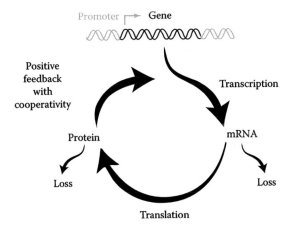

We now build a stochastic model of this circuit. The chemical species in our network include mRNA, protein, free DNA, DNA bound to one protein, and DNA bound to two proteins.

There are eight possible chemical reactions for this system:

1. Degradation of mRNA

2. Degradation of protein

3. Translation of mRNA

4. First protein binding to DNA

5. Second protein binding to DNA

6. First protein unbinding from DNA

7. Second protein unbinding from DNA

8. Transcription of DNA bound by two proteins

Note that transcription is possible only when two proteins are bound to the DNA.

Let the stochastic reaction constants be:

$$c_{mRNADegradation} = \ln(2)/120 \text{ degradations/mRNA/s}$$

$$c_{ProteinDegradation} = \ln(2)/600 \text{ degradations/protein/s}$$

$$c_{translation} = 0.167 \text{ proteins/mRNA/s}$$

$$c_{Protein1BindingDNA} = 9 \text{ associations/DNA/protein/s}$$

$$c_{Protein2BindingDNA} = 25 \text{ associations/DNA/protein/s}$$

$$c_{Protein1UnbindingDNA} = 15 \text{ dissociations/complex/s}$$

$$c_{Protein2UnbindingDNA} = 15 \text{ dissociations/complex/s}$$

$$c_{transcription} = 0.5 \text{ mRNA/s}$$

Furthermore, the initial conditions of the system are:

$$mRNA = 0 \text{ molecules}$$

$$Protein = 10 \text{ molecules}$$

$$Free\ DNA = 1 \text{ molecule}$$

$$DNA \text{ bound to one protein} = 0 \text{ molecules}$$

$$DNA \text{ bound to two proteins} = 0 \text{ molecules}$$

a. Use MATLAB to write a Gillespie algorithm to understand the time dynamics of this system. Run your algorithm for 500,000 steps. Plot the amounts of mRNA and free protein, as well as the number of proteins bound to the DNA, over time. You may consider plotting these on separate graphs if their scales are very different. For the sake of readability, you may also choose to plot fewer steps (for example, every 10th step). Describe your results in your own words.

b. Now, run the algorithm five more times. How do the simulations differ from one another? Where does the difference come from?

c. Change the parameters to make the processes and interactions in the system occur less frequently, as we would often see in individual bacterial cells. Use your Gillespie code from (a), except with these new parameter values:

$$c_{mRNADegradation} = \ln(2)/120 \text{ degradations/mRNA/s}$$

$$c_{ProteinDegradation} = \ln(2)/600 \text{ degradations/protein/s}$$

$$c_{translation} = 0.167 \text{ proteins/mRNA/s}$$

$$c_{Protein1BindingDNA} = 0.001 \text{ associations/DNA/protein/s}$$

$$c_{Protein2BindingDNA} = 0.1 \text{ associations/DNA/protein/s}$$

$$c_{Protein1UnbindingDNA} = 0.005 \text{ dissociations/complex/s}$$

$$c_{Protein2UnbindingDNA} = 0.005 \text{ dissociations/complex/s}$$

$$c_{transcription} = 0.005 \text{ mRNA/s}$$

In this case, only run your algorithm for 1,000 steps and plot the amounts of mRNA and free protein over time. Again, you may want to run your code a few times to note the trends in outcomes. Comment on any interesting patterns that you see in your graphs of mRNA and free protein. Describe any relationships that you see between the mRNA and protein quantities.

PROBLEM 6.4
Stochastic Modeling of the *mf*-Lon Network Using the Gillespie Algorithm

Let's return to the *mf*-Lon network that we considered in Problems 4.3 and 5.4, in which the expression of phage lambda protein positively regulates its own expression but can be degraded by the expression of the Lon protease. For this problem, we consider the system with a small number of molecules. As before, we assume cooperative binding ($H = 2$). We also assume that the state in which the promoter is bound by one cI molecule is a very short-lived intermediate; therefore, each DNA-binding reaction can be modeled as:

$$2(\text{cI protein}) + (\text{free } P_{RM}) \rightarrow (P_{RM} \text{ bound to 2 cI proteins})$$

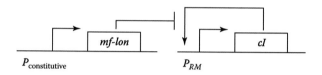

To define our stochastic model, we need to explicitly define all possible chemical reactions for each of the chemical species. The chemical species in our network include *mf*-Lon mRNA, *mf*-Lon protein, cI mRNA, cI protein, free P_{RM}, and P_{RM} bound to two cI proteins. Transcription is possible only when there are two cI proteins bound to the promoter.

There are 12 reactions in our system:

1. Transcription of *mf*-Lon mRNA

2. Degradation of *mf*-Lon mRNA

3. Translation of *mf*-Lon protein

4. Degradation of *mf*-Lon protein

5. Binding of two cI proteins to P_{RM}

6. Unbinding of two cI proteins from P_{RM}

7. Transcription of cI mRNA

8. Degradation of cI mRNA

9. "Leaky" expression of cI mRNA (not dependent on active P_{RM})

10. Translation of cI protein

11. Degradation of cI protein

12. Enzymatic degradation of cI protein by mf-Lon

Let the stochastic rate constants be:

$$c_{mf\text{-}LonTranscription} = 0.001 \text{ mRNAs/s}$$

$$c_{mf\text{-}LonRNADegradation} = c_{cImRNADegradation} = 2.38 \times 10^{-3} \text{ degradations/mRNA/s}$$

$$c_{mf\text{-}LonTranslation} = c_{cITranslation} = 5 \times 10^{-5} \text{ proteins/mRNA/s}$$

$$c_{mf\text{-}LonDegradation} = c_{cIDegradation} = 10^{-4} \text{ degradations/protein/s}$$

$$c_{2cIBindPRM} = 0.05 \text{ associations/DNA/protein}^2\text{/s}$$

$$c_{2cIUnbindPRM} = 1.25 \times 10^{-9} \text{ dissociations/complex/s}$$

$$c_{cImRNATranscription} = 1.35 \times 10^{-9} \text{ mRNAs/s}$$

$$c_{cILeakage} = 0.1 \text{ mRNAs/s}$$

$$c_{cIDegradationBymf\text{-}Lon} = 0.071 \text{ degradations/copies of cI/copies of } mf\text{-Lon/s}$$

Let the initial conditions of the system be:

Molecules of mf-Lon mRNA = 0

Molecules of mf-Lon protein = 0

Free P_{RM} = 1

P_{RM} bound to two cI proteins = 0

Molecules of cI mRNA = 0

Molecules of cI protein = 0

a. Use MATLAB to write a Gillespie algorithm to understand the time dynamics of this system. Run your algorithm for 50,000 steps. Plot the numbers of molecules of mRNA and protein for both *mf*-Lon and cI over time. Provide a brief description of the time dynamics. As your algorithm is stochastic, you may want to run your code a few more times to note the trends in outcomes.

b. Now perform an "experiment!" You can make the *mf*-Lon protein much more stable by changing the *mf*-Lon protein degradation parameter by 100-fold. What happens to the dynamic behavior of the system following this change? Provide plots of the resulting dynamics for three simulations.

c. Referring to Problems 4.3 and 5.4, comment on how the behavior you observed in your stochastic simulations versus your ODE-based analyses could inform the method you use for your modeling. Similarly, explain how the correct choice of a modeling algorithm has an impact on the prediction of the behavior of a biological system.

II

From Circuits to Networks

CHAPTER 7

Transcriptional Regulation

LEARNING OBJECTIVES

- Identify the major motifs in gene transcriptional regulatory networks

- Calculate the dynamic properties associated with these motifs using multiple methods

- Compare model outputs with corresponding experimental data

- Understand how simpler motifs combine to form larger, more complex motifs

You've seen most of the commonly used approaches to modeling biological networks, and you've applied these approaches to some relatively simple circuits. Now it's time to extend these techniques to larger, more complicated biological systems! Section II will lead you through the strategies that researchers have used to tackle three kinds of biological networks at a larger scale: transcriptional regulation, or the modification of transcription factor activity to affect gene expression; **signal transduction**, or how cells sense their surrounding environments and initiate appropriate responses; and carbon-energy **metabolism**, which breaks down nutrients from the environment to produce all of the building blocks to make a new cell.

To address these topics, I'll have to cover much more biology. You'll also learn another analysis method or two along the way, but by and large, we will be applying methods that you have already learned to problems that are more difficult. You will soon find that you are already well equipped to understand and critique existing models as well as to create your own!

TRANSCRIPTIONAL REGULATION AND COMPLEXITY

As an example, let's consider gene expression again. We've already looked at regulation extensively in Section I, but this chapter is going to add a new layer of complexity, including the interactions of multiple transcription factors to produce more complex expression dynamics.

Let's start with one of the most exciting events in the history of biology: the publication of the human genome in 2001. David Baltimore, a preeminent biologist and Nobel laureate, commented on the event: "I've seen a lot of exciting biology emerge over the past 40 years. But chills still ran down my spine when I first read the paper that describes the outline of our genome" (Baltimore, 2001; reprinted by permission from Macmillan Publishers Ltd.: *Nature*, 2001).

At the time, I remember that a number of interesting aspects of the sequence had us talking. As Baltimore wrote: "What interested me most about the genome? The number of genes is high on the list. ... It is clear that we do not gain our undoubted complexity over worms and plants by using many more genes. Understanding what does give us our complexity ... remains a challenge for the future" (Baltimore, 2001; reprinted by permission from Macmillan Publishers Ltd.: *Nature*, 2001).

It's not only the number of the genes in these genomes that was surprising; many genes encode proteins that are essentially the same and carry out the same functions, even in different organisms. Many scientists had previously assumed that the differences between species depended mostly on different genes: A human had human genes, a mouse had mouse genes, a fish had fish genes, and a sea urchin had sea urchin genes. However, the genomic sequences of all of these organisms suggested that differences in the gene complement played a much smaller role than first anticipated. For example, we share nearly all of our genes with mice.

So, what makes us different? The key is not primarily in the genes themselves, but how they are expressed. Humans have approximately eight times as much DNA sequence as the puffer fish, but essentially the same number of genes. That "extra" DNA used to be called "junk" (honestly!), but it's now clear that most of it is functional, and one of its main functions appears to be the regulation of gene expression.

Gene expression can be regulated at several points, including transcription of DNA to RNA; RNA processing, localization, and degradation; translation of the mRNA transcript into a peptide chain; and activity of the final protein. This chapter concentrates on transcriptional regulation (protein activity is discussed in the next chapters). As you will remember from Figure 1.2, we introduced this control with the example of a transcription factor that is bound by a small molecule, changing the protein's affinity for its binding site on the DNA. The binding of the transcription factor in turn affects the recruitment of the RNA polymerase complex and, subsequently, the expression of mRNA from the gene.

MORE COMPLEX TRANSCRIPTIONAL CIRCUITS

Figure 1.2 contained the simplest example we could have considered; now let's move toward more complex modeling of transcriptional regulation by considering a pair of transcription factors acting on the same gene promoter. The classic real-world example of such regulation is the regulation of the *lac* genes, whose gene products enable *E. coli* to grow on lactose and whose expression depends on two transcription factors shown in Figure 7.1a. *E. coli* prefers to eat glucose, and it will not metabolize anything else until the glucose is gone. In *E. coli*, this metabolic switching is accomplished with the transcription factor CRP (you encountered CRP in Figure 1.1). CRP binds the promoters of hundreds of genes once it is bound to cyclic adenosine monophosphate (cAMP), a small molecule whose presence indicates that none of *E. coli*'s favorite sugar sources are available. The CRP-cAMP complex can then bind operator sites that control the expression of genes that enable the uptake and metabolism of other carbon sources.

One of these carbon sources is lactose, a sugar characterized by a specific bond between galactose and glucose. Utilization of lactose depends on enzymes and a transporter, the genes for which appear together on the *E. coli* chromosome as a single transcription unit—an operon. The transcription of multiple genes as a single transcript is a more efficient way for bacteria to coordinate gene expression. Operons are common in bacteria but are largely absent from more complex organisms, which tend to rely on complex post-transcriptional regulatory processes.

The promoter of the *lac* operon contains a binding site for the CRP-cAMP complex. Thus, the operon is only fully expressed in the presence

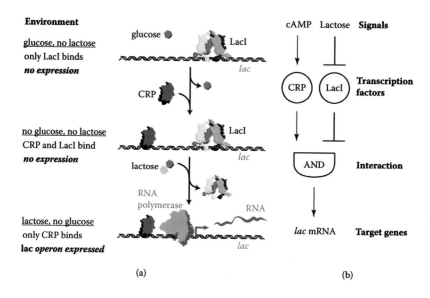

FIGURE 7.1 The *E. coli* transcription factors CRP and LacI interact on the *lac* promoter to control gene expression. (a) Schematic of the various inputs to the *lac* promoter, the combinations of which determine the transcriptional state of the *lac* operon. (b) The same system represented as a circuit diagram. Pointed arrowheads denote positive regulation; blunt arrowheads indicate negative regulation.

of CRP-cAMP, which only appears in the absence of glucose. However, what if there is no lactose in the environment? It does not make sense to express the *lac* genes unless both glucose is absent *and* lactose is present. *E. coli* addresses this problem with another transcription factor: LacI (the "I" stands for "inhibitor"). Free LacI inhibits transcription by binding its own operator in the *lac* operon promoter. However, when lactose is present in the external environment, one of its metabolic products (allolactose) binds LacI, reducing LacI's binding affinity to the operator and enabling the transcription of the *lac* operon.

So, now you know that there are two interactions between transcription factors and operator sites upstream of the *lac* operon (Figure 7.1a). However, this information is not sufficient to predict the transcriptional state of the *lac* operon because you also have to know how these two sites interact. In our case, CRP-cAMP has to be bound to the DNA *and* LacI must be absent from the DNA for maximum expression—an AND relationship. In other cases when two transcription factors regulate expression of the same transcription unit, either one of the transcription factors may be sufficient to induce expression—an OR relationship.

The biology that described in detail here is sometimes represented more compactly as illustrated in Figure 7.1b. Signals are shown as activating (arrow) or inhibiting (blunt arrow) the activity of transcription factors, which then interact at the chromosome to determine the expression of target genes. Based on this diagram, we can use a simple function (CRP AND NOT LacI) to represent the interaction, but the representation of regulatory control does not have to be Boolean.

As you can imagine, the regulation of transcription can be much more complicated in mammalian systems, with some promoters more than 10 kb long (~100 times longer than the typical bacterial promoter) and significantly longer than the gene itself. There are also many more factors that interact at the promoter, ~30 on average in some systems. We consider an example involving a mammalian transcription factor in Chapter 8.

THE TRANSCRIPTIONAL REGULATORY FEED-FORWARD MOTIF

To move toward such complex networks, let's discuss more advanced regulation architectures in *E. coli*. Previously, we examined a representation of the *E. coli* transcriptional regulatory network (Figure 1.1), and we spent a considerable amount of time studying the most common motif of that network: simple autoregulation, by which the protein product of a given gene regulates the expression of that gene. This motif was the only significant motif that was identified in the regulatory network when a single gene was considered; when two genes were considered, no additional motifs of significance were found (see Rosenfeld, Elowitz, and Alon, 2002).

Consideration of three-gene combinations, however, revealed an interesting motif (Figure 7.2). Here, transcription factor X regulates the expression of two other proteins, Y and Z. However, Y is also a transcription factor, and X and Y both regulate the expression of Z. This motif was called a **feed-forward loop**. (Control theorists: remain calm! This utilization of "feed-forward" does not correspond well to the usage in control theory. That's okay.)

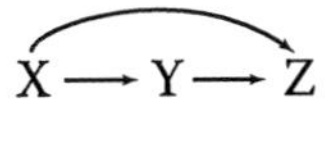

$$X \longrightarrow Y \longrightarrow Z$$

FIGURE 7.2 The feed-forward loop. This motif was found much more frequently in biological transcription networks than in randomly generated networks. The transcription factor X regulates the expression of two proteins, Y and Z. However, Y also regulates Z.

This motif occurred far more often in the *E. coli* regulatory network than would be expected by random chance, as was demonstrated by performing the same analysis on randomized networks that had the same nodes and the same number of connections as the *E. coli* network, but the connections were randomly scrambled. In the randomized networks, an average of only about two feed-forward motifs were identified, but in the real *E. coli* network, there were 42 (Douglas Adams, are you reading this?!?).

What makes feed-forward loops special? The short answer is that they exhibit fascinating and useful dynamics in terms of how they express their genes. To arrive at this answer, we first have to recognize that there are many kinds of feed-forward loops. Figure 7.3 illustrates all eight of the possible combinations of negative and positive regulation. Each combination has a direct arm from X to Z (top) and an indirect arm through Y (bottom), and either arm can exert a positive or negative influence on the expression of Z.

Now, notice that for some of the combinations, both arms exhibit the same kind of influence, whether positive or negative. For other combinations, the arms have different influences: one arm regulates positively and the other negatively. When the direct and indirect arms regulate in the same way, the combination is called **internally consistent** (or sometimes "coherent"), meaning that the direct regulation from X to Z matches the indirect regulation through Y. For example, on the upper-left loop in Figure 7.3, X has a positive and direct effect on Z, but it also exerts a positive and indirect effect because X positively regulates Y, and Y positively regulates Z. Keep in mind

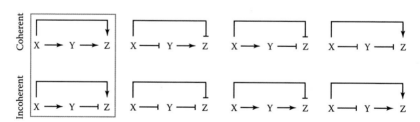

FIGURE 7.3   Possible instances of regulatory feed-forward loops. Both arms of internally consistent feed-forward loops exert the same type of control, whether positive or negative. Internally inconsistent loops have one positive arm and one negative arm. As in Figure 7.2, pointed arrowheads represent positive regulation, and blunt arrowheads denote negative regulation. The red box highlights the two instances that were specifically overrepresented in both *E. coli* and yeast. (From Alon, U. *An Introduction to Systems Biology: Design Principles of Biological Circuits.* Boca Raton, FL: Chapman and Hall/CRC, 2007. Reproduced with permission of Taylor & Francis Group LLC in the format Republish in a book via Copyright Clearance Center.)

that if Y has a negative effect on Z and X has a negative effect on Y, then the overall indirect effect of X on Z is positive. Four of the feed-forward loop combinations are internally consistent, and the other half are internally inconsistent; the direct and indirect actions of X on Z are opposite.

BOOLEAN ANALYSIS OF THE MOST COMMON INTERNALLY CONSISTENT FEED-FORWARD MOTIF IDENTIFIED IN *E. COLI*

Going back to the *E. coli* network, only two of the feed-forward loop submotifs, one coherent and one incoherent, were actually found to significantly contribute to the regulatory network (red box in Figure 7.3). First, let's consider the coherent feed-forward loop (Figure 7.4). Here, a gene g_x is transcribed to produce a protein p_x, which can then be activated by a signal s_x. The active protein p_x^* can bind the promoters for genes g_y and g_z. The product of g_y is p_y, which can be activated by s_y to produce p_y^*. The combination of p_x^* and p_y^* at the g_z promoter leads to the production of p_z.

Let's look at the dynamics of this circuit. A good place to start is with a Boolean analysis, similar to Chapter 2. Our inputs will be s_x and s_y, and we assume that all of the genes are present ($g_x = g_y = g_z = 1$). Furthermore, as in previous chapters, assume that sufficient activation of p_x and p_y occurs essentially instantaneously if s_x and s_y are present. Our equations are then reduced to:

$$\text{Activation}_{px} = \text{IF } (s_x)$$
$$p_x^* = \text{IF } (\text{Activation}_{px}) \text{ AFTER SOME TIME}$$
$$\text{Activation}_{py} = \text{IF } (p_x^*) \text{ AND } (s_y)$$
$$p_y^* = \text{IF } (\text{Activation}_{py}) \text{ AFTER SOME TIME}$$
$$\text{Expression}_{pz} = \text{IF } (p_x^*) \text{ AND } (p_y^*)$$
$$p_z = \text{IF } (\text{Expression}_{pz}) \text{ AFTER SOME TIME}$$

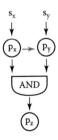

FIGURE 7.4   Gene regulatory circuit diagram for the most common internally consistent feed-forward loop in *E. coli*. Notation is as in Figure 7.1b, but notice the arrow (red), which highlights the positive regulation of p_y by p_x.

Using these equations, we can draw the (partial) state diagram shown in Figure 7.5. We begin with initial conditions of no active protein p_x^*, p_y^*, or expressed protein p_z, and sudden addition of signals s_x and s_y. Addition of s_x leads to the activation of p_x to p_x^*, followed by the expression and activation of p_y^* (expression is the step that takes more time; both processes are lumped into our equations). Finally, once p_x^* and p_y^* are active, p_z is expressed to yield the steady state of the circuit.

The dynamics of this circuit appear in Figure 7.5b. Notice that there are two periods of expression between activation of the circuit and the expression of p_z. If the circuit were a simple induction circuit with no feedback (if only p_x^* regulated the expression of p_z), then p_z would be expressed in roughly one-half the time that it takes with the feed-forward motif. The motif therefore increases the time required for expression of the target gene.

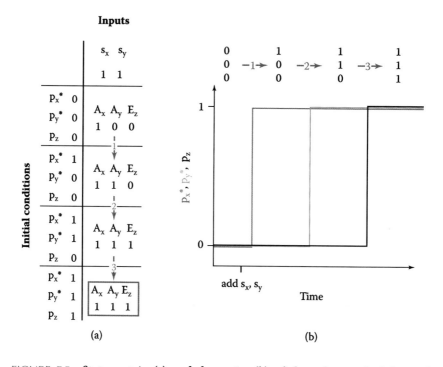

(a)

(b)

FIGURE 7.5 State matrix (a) and dynamics (b) of the coherent feed-forward loop in Figure 7.4 in the presence of both signals. After p_x is activated to p_x^*, p_y can be expressed and subsequently activated, after which p_z can be expressed. Note that the increase in p_x^* occurs after stimulus is added because the Boolean rule is AFTER SOME TIME. The numbers and arrows at the top of (b) relate to the values of p_x^*, p_y^*, and p_z over time. A, activation; E, expression.

Figure 7.5 is our analysis of what happens when the circuit is initially inactive (no stimuli) and we then activate it; contrast that behavior with a circuit that is initially active and is then inactivated. Begin with the state-matrix approach shown in Figure 7.6a. When the stimulus is removed, p_x^* becomes rapidly deactivated to p_x. The expression of both p_y and p_z depends on p_x^*, so neither protein can be produced. As a result, both proteins decay over the same time period (Figure 7.6b).

Thus, when this system is activated, there are two time steps: one for the expression and activation of p_y^* and one for the expression of p_z. However, when the system is deactivated, there is only a single time step because p_y^* and p_z are removed simultaneously. This coherent feed-forward loop created a switch with different ON and OFF times. Why is this strategy useful? Uri Alon, whose team at the Weizmann Institute of Science originally identified these motifs, gave the example of an elevator door. It is important that the door be safe, which means that it should start to close slowly but stop closing quickly, for example, at the instant that someone's foot triggers the safety mechanism.

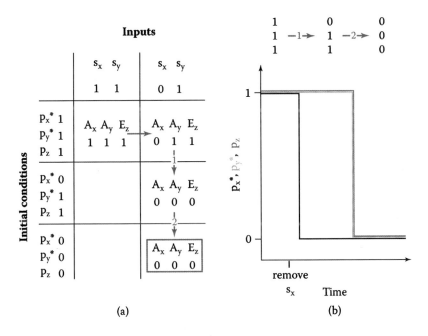

(a)

(b)

FIGURE 7.6 State matrix (a) and the dynamics (b) illustrate the deactivation of the feed-forward loop in Figure 7.4. In this case, s_x is switched to zero, which leads to deactivation of p_x^* (red arrow 1), followed by a halt in the expression of p_y and p_z. Compare these figures with Figure 7.5.

AN ODE-BASED APPROACH TO ANALYZING THE COHERENT FEED-FORWARD LOOP

We have discussed how to analyze this circuit using a Boolean approach; a parallel approach (favored by Alon, 2007) would be to write ODEs as presented next. We begin with $d[p_y]/dt$ in the usual formulation:

$$\frac{d[p_y]}{dt} = prod - loss \tag{7.1}$$

We still represent the loss term as proportional to the amount of $d[p_y]/dt$. For production, we will invoke a **threshold concentration**: Transcription of p_y can only begin once $[p_x^*]$ reaches a certain value. We indicate threshold notations with the notation K_{ab}, where a denotes the transcription factor and b the target gene; therefore, we will add a K_{xy} term to Equation 7.1. We further assume that the expression of p_y is maximal when $[p_x^*]$ is greater than the threshold constant K_{xy}, and that otherwise the expression of p_y is equal to zero. Adding these details to our equation, we obtain:

$$\frac{d[p_y]}{dt} = k_{yprod} \cdot \theta\left([p_x^*] > K_{xy}\right) - k_{ydeg}[p_y] \tag{7.2}$$

where the function $\theta(statement)$ is equal to one if the statement in parentheses is true and equal to zero if the statement is false. For the case in which s_x has been added at a sufficient concentration for activation, $\theta = 1$, and Equation 7.2 is reduced to:

$$\frac{d[p_y]}{dt} = k_{yprod} - k_{ydeg}[p_y] \tag{7.3}$$

We've solved equations like this before (in Chapter 3, for example), so you should be able to show that the solution of this equation is:

$$[p_y](t) = [p_y]_{ss}\left(1 - e^{-k_{ydeg}t}\right) \tag{7.4}$$

where $[p_y]_{ss}$ is determined from Equation 7.3:

$$0 = k_{yprod} - k_{ydeg}[p_y]_{ss} \tag{7.5}$$

and therefore,

$$[p_y]_{ss} = \frac{k_{yprod}}{k_{ydeg}} \qquad (7.6)$$

In general, the reactions that lead to activation of a transcription factor happen at a significantly faster rate (on the order of seconds or less) than the rate of gene expression (minutes). As a result, for our purposes here we assume that the transition from p_y to p_y^* is very fast in the presence of s_y, so in this case, $[p_y](t) = [p_y^*](t)$.

The equation for p_z is similar to the equation for p_y, but in this case, the production of p_z is based on two conditions occurring simultaneously:

$$\frac{d[p_z]}{dt} = k_{zprod}\,\theta\!\left([p_x^*] > K_{xz}\right) \cdot \theta\!\left([p_y^*] > K_{yz}\right) - k_{zdeg}[p_z] \qquad (7.7)$$

When $[p_x^*]$ and $[p_y^*]$ are sufficiently large, the solution is similar to that for $[p_y]$:

$$[p_z](t) = [p_z]_{ss}\left(1 - e^{-k_{zdeg}t}\right), \quad [p_z]_{ss} = \frac{k_{zprod}}{k_{zdeg}} \qquad (7.8)$$

The response of the system to a sudden addition of s_x and s_y is depicted in Figure 7.7. After s_x is added, the expression of p_y and consequently p_y^* increases until $[p_y^*]$ reaches the threshold K_{yz}. At this time, p_z expression is induced.

Let's consider the time required for $[p_y^*]$ to reach the threshold K_{yz}, as this delay was found to be the most notable aspect of this circuit's dynamic response. As shown in Figure 7.8, we call this delay T_{on} and determine it from Equation 7.4, remembering that $[p_y](t) = [p_y^*](t)$ if s_y is present:

$$[p_y^*](t = T_{on}) = [p_y]_{ss}\left(1 - e^{-k_{ydeg}T_{on}}\right) \qquad (7.9)$$

Since $[p_y^*](t = T_{on}) = K_{yz}$, we can solve for T_{on}:

$$K_{yz} = [p_y]_{ss}\left(1 - e^{-k_{ydeg}T_{on}}\right) \qquad (7.10)$$

$$T_{on} = \frac{1}{k_{ydeg}} \ln\!\left(\frac{1}{1 - \dfrac{K_{yz}}{[p_y]_{ss}}}\right) \qquad (7.11)$$

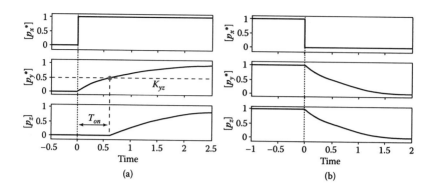

FIGURE 7.7 Dynamic response of the coherent feed-forward loop in Figure 7.4 to (a) sudden addition or (b) sudden removal of s_x at time zero. Notice that there is a delay (T_{on}) between expression changes in p_y* and p_z when s_x is added, but not when s_x is removed, as we also observed in our Boolean analysis (Figures 7.5 and 7.6). Remember that $[p_y] = [p_y*]$ because of an assumption of rapid p_y activation. (Modified from Alon, U. *An Introduction to Systems Biology: Design Principles of Biological Circuits*. Boca Raton, FL: Chapman and Hall/CRC, 2007. Reproduced with permission of Taylor & Francis Group LLC in the format Republish in a book via Copyright Clearance Center.)

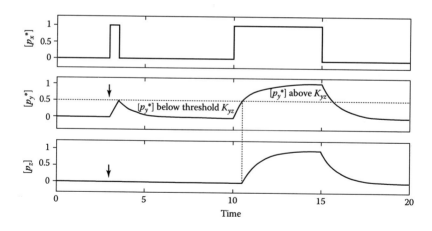

FIGURE 7.8 Response of the feed-forward loop in Figure 7.4 to a brief pulse of s_x. The pulse of stimulus (highlighted by an arrow in the bottom plots) leads to only brief activation of p_x, which in turn leads to a maximum expression of p_y that is below the threshold K_{yz} (dashed line). As a result, p_z expression is never induced. For contrast, a longer pulse of s_x and its consequences are shown at right (and in Figure 7.7a). (Adapted by permission from Macmillan Publishers Ltd.: Shen-Orr, S. S., Milo, R., and Alon, U. *Nature Genetics* 2002, **31**(1): 64–68.)

From Equation 7.11, you can see that when $[p_{y,ss}]$ is much larger than K_{yz}, $K_{yz}/[p_{y,ss}]$ approaches zero, which means that T_{on} will also reduce to zero for any k_{ydeg}. As the value of $[p_{y,ss}]$ approaches the threshold value K_{yz}, the logarithm term of Equation 7.11 increases rapidly, meaning that the value of T_{on} increases dramatically as well.

ROBUSTNESS OF THE COHERENT FEED-FORWARD LOOP

As you can see, the overall conclusions that we drew in the previous section using ODEs were similar to the results of our Boolean-based approach in the section on analyzing the coherent feed-forward loop. However, ODE-based formulation of the model also allows us to demonstrate another interesting aspect of this coherent feed-forward circuit: its **robustness**. In this context, robustness means that the system will not change much in response to a small perturbation. As an example, let's say that our circuit is exposed only briefly to s_x. As before, p_y begins to be expressed, but it does not reach the critical threshold for p_z expression; thus, p_z is never expressed (Figure 7.8). The robustness of the system is tuned by the value of the threshold: A higher threshold takes longer for the p_y value to attain, so expression of p_z would be robust to even longer stimulus times.

EXPERIMENTAL INTERROGATION OF THE COHERENT FEED-FORWARD LOOP

Having mathematically analyzed this coherent feed-forward circuit in detail, Alon's team decided to see whether the experimental data from a naturally occurring feed-forward circuit in *E. coli* actually exhibited the dynamics that theory predicted. They focused on the *ara* genes, which are regulated by the transcription factors AraC and CRP. AraC is regulated transcriptionally by CRP and is activated in the presence of arabinose (Figure 7.9b). By keeping the bacteria in an arabinose environment and suddenly adding or "removing" cAMP (see Figure 7.9 caption), the team replicated the situation that they modeled. To enable comparison, the team simultaneously considered a control circuit that responded directly to CRP but not to AraC (gray lines in Figures 7.9b and 7.9c). Consistent with the theory shown in Figure 7.7, expression of the *ara* genes was delayed when cAMP was added (Figure 7.9b), but not when cAMP was removed (Figure 7.9c)—an elegant demonstration of their theory!

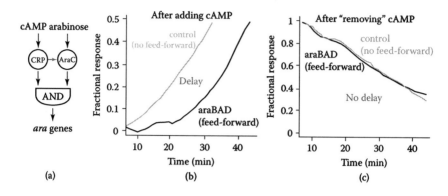

FIGURE 7.9 Monitoring promoter states when turning on and off circuits with (black) and without (gray) feed-forward components. The feed-forward circuit (a) and a control circuit (based on the *lac* operon) were (b) turned on by adding saturating amounts of cAMP to growing cells and (c) turned off by "removing" cAMP (actually by adding saturating glucose, which inactivates CRP). The promoters were transcriptionally fused to GFP, which acted as a reporter of promoter activity. Note that turning on the circuit is delayed in the feed-forward loop, but the two circuits have similar dynamics when turning off. (From Mangan, S., Zaslaver, A., and Alon, U. *Journal of Molecular Biology* 2003, **334**(2): 197–204. Reprinted with minor modifications with permission from Elsevier.)

CHANGING THE INTERACTION FROM AN AND TO AN OR RELATIONSHIP

In our analysis of the coherent feed-forward circuit in the four previous subsections, we focused on an AND relationship between the transcription factors controlling gene expression: Both p_x^* AND p_y^* were required. Several other relationships are possible that can lead to differences in network behavior. For example, let's change our current circuit such that the interaction at the promoter changes from an AND interaction to an OR interaction. We can use our Boolean toolbox for a quick analysis. Our equations for Activation$_{px}$, p_x^*, Activation$_{py}$, p_y^*, and p_z remain the same as in the subsection on Boolean analysis of the most common internally consistent feed-forward motif identified in *E. coli*, but the equation for p_z expression becomes:

$$\text{Expression}_{pz} = \text{IF } (p_x^*) \text{ OR } (p_y^*)$$

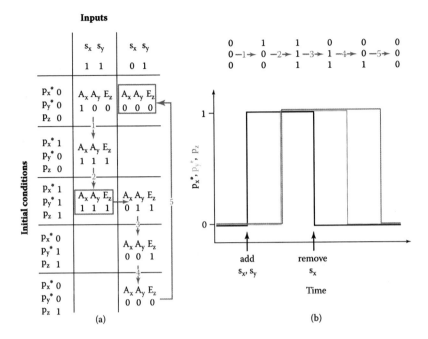

FIGURE 7.10 The state matrix (a) and dynamics (b) for the OR circuit. Note the delay when the stimulus s_x is removed, but not when it is added.

The state matrix and dynamics for the OR circuit appear in Figure 7.10. Notice that in this case, the delay occurs when the stimulus is removed. This simple change in the interaction at the promoter therefore determines whether the delay occurs in the expression of p_z or in the decay of p_z.

Again, Alon's team set out to experimentally verify these predictions. They focused their investigation on some of the genes that regulate expression of the bacterial flagellum; these genes are naturally regulated by a coherent feed-forward circuit with an OR interaction. As shown in Figure 7.11, their experimental results strongly agreed with the theory we worked through so far in this chapter.

PRACTICE PROBLEM 7.1

Now that we've analyzed a coherent feed-forward loop from *E. coli*, let's consider the primary incoherent feed-forward loop in Figure 7.12. Draw a state diagram and calculate the dynamic response for the case in which s_x and s_y are suddenly added.

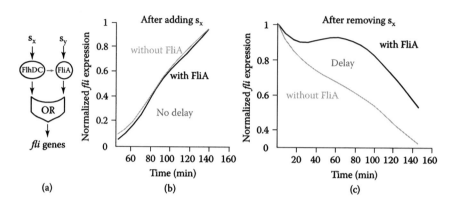

FIGURE 7.11 Experimental validation of feed-forward dynamics for the coherent feed-forward loop with an OR interaction at the promoter. (a) The *E. coli* flagellar system, a control circuit with two inputs and an OR interaction. Both of these circuit types occur naturally in *E. coli*. (b) For experimental validation, the production of FlhDC is controlled with a promoter that is induced by the addition of arabinose (not the native promoter), which serves as s_x. The signal s_y comes from a checkpoint system that monitors the production of a component of the flagellum. The promoter controlling the *fli* sequences at the bottom of the circuit is fused to GFP as a reporter. Here, the "on" step is similar to that for a feed-forward circuit, which occurs when FliA is deleted. (c) Turning the circuit off (by shifting the cells into medium without arabinose) is delayed for the feed-forward circuit when FliA is deleted (note the similarity to turning off the circuit in Figure 7.9). (Adapted with permission from Macmillan Publishers Ltd.: Kalir, S., Mangan, S., and Alon, U. *Molecular Systems Biology*. Epub 2005 Mar 29. doi: 10.1038/msb4100010.)

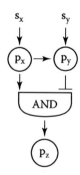

FIGURE 7.12 An incoherent feed-forward loop, in which p_x exerts a positive influence on p_y and p_z, but p_y has a negative influence on p_z.

SOLUTION

The regulatory rules for this feed-forward loop can be written as:

$$\text{Activation}_{px} = \text{IF } (s_x)$$
$$p_x{}^* = \text{IF } (\text{Activation}_{px}) \text{ AFTER SOME TIME}$$
$$\text{Activation}_{py} = \text{IF } (p_x{}^*) \text{ AND } (s_y)$$
$$p_y{}^* = \text{IF } (\text{Activation}_{py}) \text{ AFTER SOME TIME}$$
$$\text{Expression}_{pz} = \text{IF } (p_x{}^*) \text{ AND NOT } (p_y{}^*)$$
$$p_z = \text{IF } (\text{Expression}_{pz}) \text{ AFTER SOME TIME}$$

Using these rules, we draw a state matrix (Figure 7.13a) and plot the resulting dynamics (Figure 7.13b).

This circuit creates a "pulse" of p_z expression! You can imagine that this response is useful to the cell. When real instances of these feed-forward loops were examined in *E. coli*, a circuit was identified (Figure 7.14a) in which $[p_z]$ decreased to a new, non-zero steady-state level predicted by the ODEs (Figure 7.14b); experimental data confirmed this prediction (Figure 7.14c).

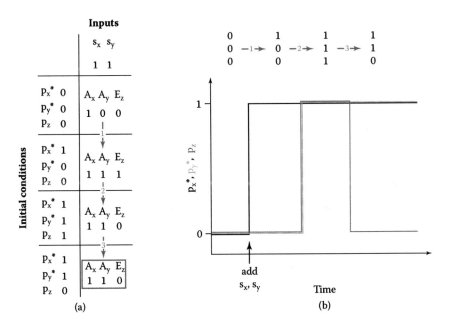

FIGURE 7.13 The state matrix (a) and dynamics (b) for the incoherent feed-forward loop in Figure 7.12. Notice that the expression of p_z rises, then falls again, in a pulse.

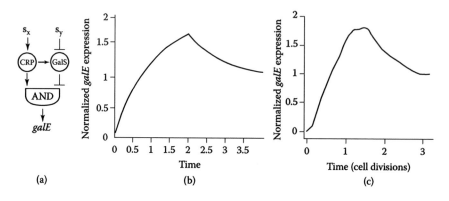

FIGURE 7.14 Dynamics of the incoherent feed-forward loop (a) as determined using ODEs (b) and experimentally (c). Here, the *galE* promoter is fused to GFP as a reporter. s_x is cAMP, which activates CRP, and s_y is galactose, which causes GalS to unbind from the *galE* promoter. (From Mangan, S., Itzkovitz, S., Zaslaver, A., and Alon, U. *Journal of Molecular Biology* 2006, **356**: 1073–1081. Reprinted with minor modifications with permission from Elsevier.)

THE SINGLE-INPUT MODULE

The feed-forward loop was the only motif that Alon's team found in all of the possible three-node interaction sets (Figure 7.3). Now let's talk about some of the other motifs that were found when considering greater numbers of nodes, which are easy to describe but exhibit behaviors that are a bit more complex.

A common motif, the **single-input module,** has one regulator that is solely responsible for the regulation of several genes, often including itself. The genes may be located in the same operon or the genes may be spread across the genome, in which case the genes are said to be in the same **regulon** (meaning that they are regulated by the same transcription factor but are not necessarily in physical proximity to the gene encoding that transcription factor). Single-input modules are very unlikely to occur in a random network, especially in the case of one transcription factor regulating >10 genes.

You can imagine that genes in the same regulon have coordinated expression, just as genes in operons do. However, all of the genes in a given regulon are not necessarily expressed at the same time in a single transcript, as is the general rule for genes in an operon.

For example, consider the arginine biosynthesis single-input module in *E. coli* (Figure 7.15). These genes encode the enzymes required to synthesize the amino acid arginine, and they are only expressed when arginine is

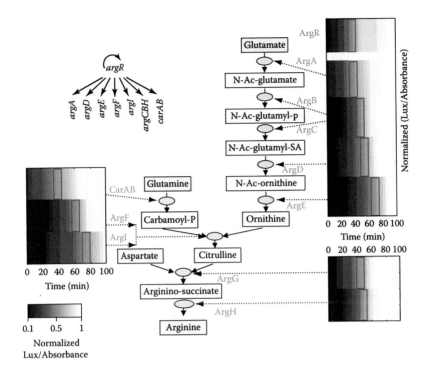

FIGURE 7.15 The arginine biosynthesis single-input motif and just-in-time expression. At the top left is the transcriptional regulatory motif in which the *argR* gene product controls its own expression as well as that of several genes that encode metabolic enzymes. These enzymes are arranged into the metabolic pathways that are responsible for arginine biosynthesis (metabolite names appear in boxes; enzymes that catalyze the conversion of one metabolite to another are represented by circles and solid arrows). The expression of each enzyme is shown in grayscale, with the lightest gray (boxed in red) indicating the inflection point at which expression of the gene reached half of its maximum. Comparing these inflection points can give you an idea of the order of gene expression. The Lux protein, which is luminescent, was used here as the reporter of promoter activity. (Adapted by permission from Macmillan Publishers Ltd.: Zaslaver, A., Mayo, A. E., Rosenberg, R., Bashkin, P., Sberro, H., Tsalyuk, M., Surette, M. G., and Alon, U. *Nature Genetics* 2004, **36**(5): 486–491.)

not present in *E. coli*'s environment. The genes fall into seven transcription units: the five individually expressed genes *argA*, *argD*, *argE*, *argF*, and *argI*; the operon encoding *argC*, *argB*, and *argH*; and the operon encoding *carA* and *carB*. All of these genes are regulated solely by the transcription factor ArgR, which is why they belong to the same single-input module. The gene encoding ArgR is also in the regulon because it regulates its own expression.

However, *argG* is technically not in the regulon, even though its expression is regulated by ArgR, because CRP is known to also control the expression of *argG*.

Figure 7.15 illustrates the relationship between the genes in the single-input module, both in terms of transcriptional regulation and the metabolic pathway (boxes are metabolites, circles are enzymes). The figure also includes a time course detailing the expression of each gene when arginine is removed from *E. coli*'s surroundings.

JUST-IN-TIME GENE EXPRESSION

Interestingly, in Figure 7.15, the genes appear to be expressed roughly in the order they are needed, or **just in time**. For example, the enzymes encoded by *argB* and *argC* are induced just after the expression of *argA*, which catalyzes the step in the biosynthetic pathway immediately preceding them. This strategy seems to be an efficient way to ensure that the enzymes are only made when absolutely required: in the absence of arginine. For a single-celled organism like *E. coli*, the energy and space required to produce and store extra proteins are extremely valuable, so this efficiency, small as it seems, could yield important dividends in terms of growth rate.

How is just-in-time expression achieved? Let's consider a transcription factor p_x that regulates the expression of three genes: g_a, g_b, and (you guessed it) g_c. We use the same ODE-based framework as in the section entitled "An ODE-based Approach to Analyzing the Coherent Feed-forward Loop"; as previously, we assume that once s_x is added, all of the p_x is rapidly activated to $p_x{}^*$. The equations for gene expression are:

$$\frac{d\left[p_x\right]}{dt} = k_{xprod} - k_{xdeg}\left[p_x\right] \tag{7.12}$$

$$\frac{d\left[p_a\right]}{dt} = k_{aprod}\,\Theta\!\left(\left[p_x^*\right] > K_{xa}\right) - k_{adeg}\left[p_a\right] \tag{7.13}$$

$$\frac{d\left[p_b\right]}{dt} = k_{bprod}\,\Theta\!\left(\left[p_x^*\right] > K_{xb}\right) - k_{bdeg}\left[p_b\right] \tag{7.14}$$

$$\frac{d\left[p_c\right]}{dt} = k_{cprod}\,\Theta\!\left(\left[p_x^*\right] > K_{xc}\right) - k_{cdeg}\left[p_c\right] \tag{7.15}$$

The equations are identical in form, and as you already know, the production and decay terms determine the steady-state expression level of protein once the genes are induced. The thresholds K_{xa}, K_{xb}, and K_{xc} determine how long it takes for gene expression to be induced after p_x is expressed. For example, if $K_{xa} < K_{xb}$, then it will take longer for $[p_x^*]$ to be greater than K_{xb}, so p_b expression will not be induced until later. Similarly, you can see in Figure 7.16 that if $K_{xa} < K_{xb} < K_{xc}$, expression of these genes will occur in the order g_a, then g_b, then g_c—just as with the arginine regulon (Figure 7.15).

What does it mean biologically for these thresholds to be different from each other? Most likely, the difference is related to the binding affinity of the transcription factor for each gene's promoter. A promoter with a high-affinity binding site would lead to a low threshold: Even at a low concentration of the transcription factor, binding occurs and expression is induced.

The single-input module yields therefore has the potential for sequential induction of gene expression. Notice from Figure 7.16, however, that when expression of p_x decays, g_c is the first gene to respond, followed by g_b and finally g_a. In other words, deactivation of the system occurs in reverse order: **first in, last out.**

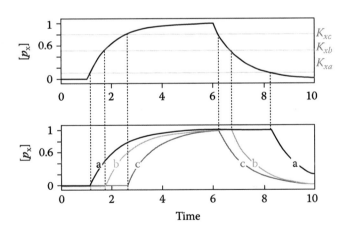

FIGURE 7.16 The temporal program of expression from a single-input module. As the activity of p_x rises, it crosses the thresholds of activity for each promoter (K_{xa}, K_{xb}, and K_{xc}) in order. When the activity of p_x declines, it exhibits first-in, last-out behavior. (Adapted by permission from Macmillan Publishers Ltd.: Shen-Orr, S. S., Milo, R., and Alon, U. *Nature Genetics* 2002, **31**(1): 64–68.

GENERALIZATION OF THE FEED-FORWARD LOOP

Are there any network structures that would give us **first-in, first-out** behavior? After all, you might expect that this strategy would be even more efficient in terms of enzyme production; enzymes are produced just in time and only maintained as long as they are needed.

As it happens, another motif that was shown to be overrepresented in *E. coli*'s network versus randomized networks can indeed produce first-in, first-out behavior. This motif is essentially a generalization of the feed-forward loop. Whereas the feed-forward loops that we considered in Figure 7.3 had only one target gene, in the generalized feed-forward loop, two transcription factors—one under the transcriptional control of the other—control the expression of many target genes. In this sense, the generalized or multigene feed-forward loop may also be conceptualized as a hybrid between the feed-forward loop and the single-input module.

PRACTICE PROBLEM 7.2

Using the ODE approach we adopted for the single-input motif, write the equations for the multigene feed-forward loop in Figure 7.17 and describe the conditions for which first-in, first-out expression of the genes *z1* and *z2* will be achieved.

SOLUTION

The equations appear below. Assume that s_x and s_y are sufficiently available such that $[p_x] = [p_x^*]$ and $[p_y] = [p_y^*]$.

$$\frac{d[p_x]}{dt} = k_{xprod} - k_{xdeg}[p_x]$$

$$\frac{d[p_y]}{dt} = k_{yprod}\,\theta([p_x^*] > K_{xy}) - k_{ydeg}[p_y]$$

$$\frac{d[p_{z1}]}{dt} = k_{z1prod}\left[\theta([p_x^*] > K_{xz1})\,\text{OR}\,\theta([p_y^*] > K_{yz1})\right] - k_{z1deg}[p_{z1}]$$

$$\frac{d[p_{z2}]}{dt} = k_{z2prod}\left[\theta([p_x^*] > K_{xz2})\,\text{OR}\,\theta([p_y^*] > K_{yz2})\right] - k_{z2deg}[p_{z2}]$$

From our work in the section on just-in-time gene expression, you should already have an intuition that first-in, first-out behavior depends on the threshold K parameters. If $K_{xz1} < K_{xz2}$, then p_{z1} will begin to be expressed earlier than p_{z2} (Figure 7.18).

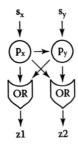

FIGURE 7.17 A two-node feed-forward loop with two gene targets. The transcription factors that control expression are common to both targets.

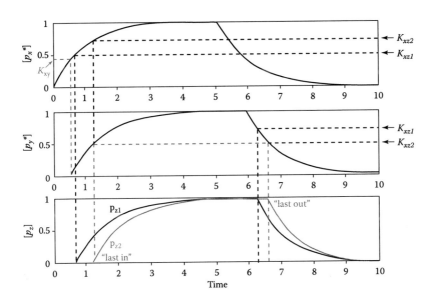

FIGURE 7.18 The output of the multinode feed-forward loop in Figure 7.17, which exhibits first-in, first-out dynamics. Due to the OR relationship between the transcription factors, the expression of p_x, which precedes p_y expression, controls the initial induction of p_{z1} (black) and p_{z2} (red). p_{z1} is expressed before p_{z2} because $K_{xz1} < K_{xz2}$. The decrease in p_y expression occurs after $[p_x]$ decreases, so $[p_y]$ controls the decreases in $[p_{z1}]$ and $[p_{z2}]$. Here, $K_{yz1} > K_{yz2}$, so p_{z1} expression decreases before p_{z2} expression does. p_{z1} levels increase first, then p_{z2} levels rise; when the levels of p_x and p_y decrease, once again p_{z1} levels decrease before p_{z2} levels do. (Modified from Alon, U. *An Introduction to Systems Biology: Design Principles of Biological Circuits.* Boca Raton, FL: Chapman and Hall/CRC, 2007. Reproduced with permission of Taylor & Francis Group LLC in the format Republish in a book via Copyright Clearance Center.)

Look at the right side of Figure 7.18 to find the requirements for "first out" or, in our case, p_{z1}'s expression dropping before that of p_{z2}. Notice that the removal of p_x from the system does not affect p_{z1} or p_{z2} due to the OR relationship that is encoded into the expression of both of the corresponding genes. Instead, it is the removal of p_y after p_x has already begun to be lost (and passed the K_{xy} threshold) that determines when p_{z1} and p_{z2} begin to be removed from the system. In this case, the requirement for p_{z1} to be removed first is reflected by the fact that $K_{yz1} > K_{yz2}$. Taken together, these parameters lead to first-in, first-out expression.

AN EXAMPLE OF A MULTIGENE FEED-FORWARD LOOP: FLAGELLAR BIOSYNTHESIS IN *E. COLI*

We have now identified the network structure (multinode feed-forward loop) and the conditions (OR interactions at the promoter and certain relationships between the threshold parameters) that lead to first-in, last-out expression of target genes. But is this kind of behavior ever exhibited by live *E. coli*? An impressive real example that produces similar dynamics occurs during construction of the bacterial flagellum, the long "tail" that *E. coli* uses to propel itself through its liquid environment.

When *E. coli* that grow flagella (some lab strains do not) are placed in an environment without much food, they use their flagella so that they can look for environments that are more nutrient rich. This system has been well characterized; we know which genes encode all of the parts of the flagellum (Figure 7.19a). As we saw with the arginine biosynthesis single-input module (Figure 7.15), the flagellar genes are expressed in roughly the order in which they are required (Figure 7.19b). However, in this case, two transcription factors regulate the expression of the flagellar biosynthesis genes: FlhDC and FliA. Furthermore, as you saw in the section entitled "Changing the Interaction from an AND to an OR Relationship," FliA expression is also regulated by FlhDC. The resulting multinode feed-forward loop is shown in Figure 7.19c.

Alon's group was particularly interested in the structure of this multigene feed-forward loop and subjected it to intense experimental scrutiny (see Kalir et al., 2001). They verified that FliA regulated the expression of even the early-expressed genes (previously thought to be only under FlhDC control), probed the nature of the interaction between FlhDC and FliA at the promoter regions (a SUM relationship, which is similar to,

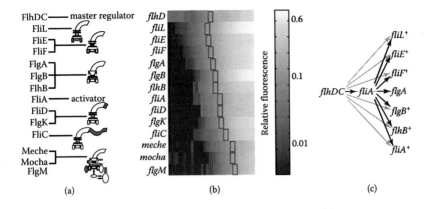

FIGURE 7.19 Regulation of the flagellar biosynthesis genes via a multinode feed-forward loop driven by FlhDC and FliA. (a) The genes are linked to the part of the flagellum that they encode. (b) Time course of gene expression for all of the genes involved in the circuit. Promoter activity was monitored by transcriptional fusions to GFP. As in Figure 7.16, locating the inflection point facilitates visualization of the order of gene expression. (c) Schematic of the regulation of the system. A plus sign indicates that only the first gene in the operon is listed. (Modified from Kalir, S., McClure, J., Pabbaraju, K., Southward, C., Ronen, M., Leibler, S., Surette, M. G., and Alon, U. *Science* 2001, **292**: 2080–2083. Reprinted with permission from AAAS.)

but slightly more complex than, an OR relationship), and determined the dynamics of activation for FlhDC, FliA, and the target transcription units.

Those dynamics can be seen in Figure 7.20. The activity of each transcription factor (corresponding roughly to $[p_x^*]$ and $[p_y^*]$ in Practice Problem 7.2) is measured by the expression of GFP, whose gene is fused to a protein that is under the sole control of either FlhDC or FliA (Figure 7.20a). The other seven promoter activities (Figure 7.20b) are taken from the endogenous flagellar biosynthesis promoters, also linked to GFP expression.

Notice that FliA activity increases sharply after FlhDC reaches its full activity. This makes sense because FliA expression depends on FlhDC. Now, look at the endogenous promoter readouts in Figure 7.20b. During the first phase of expression, when FlhDC is the predominantly active transcription factor, the promoter activities differ within an order of magnitude. We already saw in Figure 7.19b that these genes are expressed in a specific order. However, once FliA becomes the predominant transcription factor, some promoter activities increase and others decrease, such that all of the promoters have the same activity (Figure 7.20b).

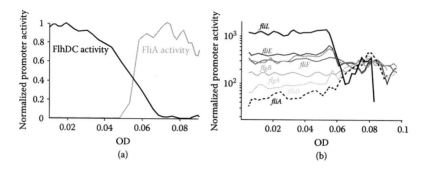

FIGURE 7.20 **Dynamics from the flagellar biosynthesis network, determined experimentally.** See Figure 7.19c for the regulatory schematic of this system. (a) Black line, the abundance of the fluorescent reporter fused to a mutated *fliL* promoter that does not bind FliA reflects the activity of the *flhDC* promoter. Gray line, fluorescent protein abundance when the reporter is fused to a promoter that is only responsive to FliA. (b) The activities of the various promoters change as the network shifts from FlhDC dominated to FliA dominated. OD stands for optical density, a measure of how much light passes through a bacterial culture; the more bacteria that have grown over time, the denser the solution. Promoter activity is normalized by dividing the measured promoter fluorescence by the OD. Normalization controls for the fact that fluorescence increases not only as a result of promoter activity, but also as a by-product of cell growth. (From Kalir, S. and Alon, U. *Cell* 2004, **117**: 713–720. Reprinted with minor modifications with permission from Elsevier.)

In other words, these activities do not quite lead to first-in, first-out expression patterns; it's more like first-in expression followed by a constant rate of maintenance across the transcription units. Such a pattern is interesting in its own right and would not be possible with a single-input module. In any event, the coordination of two transcription factors can produce dynamics that are quite complex!

OTHER REGULATORY MOTIFS

Two other motifs were found to be overrepresented in the *E. coli* and yeast transcriptional regulatory networks: the **bi-fan motif** in Figure 7.21a and the **dense overlapping region** in Figure 7.21b. The bifan motif was the only overrepresented four-node motif; it occurs when two transcription factors each regulate the same two target genes. The dense overlapping region is a generalization of the four-node motif in which a set of transcription factors regulates a common set of target genes. For this motif, it is not strictly necessary that all of the transcription factors regulate all of the

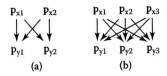

$$P_{x1} \quad P_{x2} \qquad\qquad P_{x1} \quad P_{x2} \quad P_{x3}$$

$$P_{y1} \quad P_{y2} \qquad\qquad P_{y1} \quad P_{y2} \quad P_{y3}$$

(a) \qquad\qquad\qquad (b)

FIGURE 7.21 The last two motifs found to be overrepresented in *E. coli* and yeast. (a) A bi-fan motif. (b) Dense overlapping regions.

target genes, only that all transcription factors regulate or are regulated by more than one factor or gene and all transcription factors are connected. The properties of these final two motifs have not been investigated in detail, but I mention them here for completeness. Maybe characterizing them is work for you to do someday!

Now we've covered essentially all of the motifs that have been identified in a transcriptional regulatory network. I hope you've come to appreciate how dynamics can change with network structure, and that you have obtained a sense of how you could put many of these motif models together to begin assembling networks representative of the whole cell.

CHAPTER SUMMARY

To move toward whole-cell modeling, we need to progress from the simple regulatory circuits we considered in the previous chapters of the book to more complex regulatory circuits. Motifs that occur significantly more often in the *E. coli* and yeast transcriptional regulatory networks (as compared to randomly generated networks) have been identified: autoregulation, the feed-forward loop, the single-input motif, multigene feed-forward loops, bi-fans, and dense overlapping regions.

Feed-forward loops are motifs in which a gene's expression is regulated by two transcription factors, one of which is also regulated by the other. Such loops can be internally consistent, or coherent, if both arms of the loop exert the same type of regulation (are either both positive or both negative). Transcription factors can interact in various ways to regulate gene expression; AND and OR relationships were considered and the slightly more complicated SUM relationship was mentioned.

We modeled the dynamics of three instances of feed-forward loops, primarily using the Boolean approaches developed in Chapter 2. First, we modeled a coherent feed-forward loop in which all regulation was positive and there was an AND relationship between the transcription factors. Expression of the target gene was delayed (relative to the expression of the regulated transcription factor) during induction, but not during the repression of expression. A similar feed-forward loop (only differing by an OR interaction) exhibited a delay in expression change during repression but not induction. We used the analogy of controlling the opening and closing of an elevator door to suggest how such direction-sensitive delays could be useful to the cell (such utility has not been established to date). Using ODEs, we derived an equation for the delay time and showed that these feed-forward loops also exhibit robustness to short perturbations. The third instance of a feed-forward loop was incoherent: One arm regulated target gene expression positively and the other negatively. The most common type of incoherent feed-forward loop in *E. coli* produces short pulses of target gene expression. Satisfyingly, examples of all three of these feed-forward loops have been investigated in *E. coli* and exhibit the dynamics predicted by theory.

The single-input motif has a single transcription factor that regulates the expression of many transcription units, often including its own gene. Using an ODE-based approach, we demonstrated that such networks can exhibit just-in-time dynamics, in which gene expression is induced sequentially rather than simultaneously. Just-in-time expression dynamics have been observed in *E. coli* in metabolic biosynthesis pathways, where it could be advantageous to produce the next metabolite in the pathway at sufficient concentrations before expressing the enzyme that can bind and convert it to something else.

According to our simple model, the first protein expressed in the single-input motif is destined to be the last protein remaining after repression of expression. This first-in, last-out expression may be useful in some cases, but intuition suggests that first-in, first-out dynamics, which can be displayed by multinode feed-forward loops under certain conditions, may be more useful to the cell. Although first-in, first-out dynamics have not been observed to my knowledge, we did consider a related case in the *E. coli* flagellar biosynthesis transcriptional regulatory network.

RECOMMENDED READING

Alon, U. *An Introduction to Systems Biology: Design Principles of Biological Circuits.* Boca Raton, FL: Chapman and Hall/CRC Press, 2007.

Baltimore, D. Our genome unveiled. *Nature* 2001, **409**: 814–816.

Davidson, E. H. *The Regulatory Genome: Gene Regulatory Networks in Development and Evolution.* New York: Academic Press, 2006.

Human Genome Project Information Archive. Home page. http://web.ornl.gov/sci/techresources/Human_Genome/home.shtml.

    This website, from the Oak Ridge National Laboratory and the Department of Energy, contains information about the Human Genome Project and associated issues spanning bioethics, medicine, education, and progress after the publication of the human genome sequence.

Kalir, S. and Alon, U. Using a quantitative blueprint to reprogram the dynamics of the flagella gene network. *Cell* 2004, **117**: 713–720.

Kalir, S., Mangan, S., and Alon, U. A coherent feed-forward loop with a SUM input function prolongs flagella expression in *Escherichia coli*. *Molecular Systems Biology* Epub 2005 Mar 29. doi: 10.1038/msb4100010.

Kalir, S., McClure, J., Pabbaraju, K., Southward, C., Ronen, M., Leibler, S., Surette, M. G., and Alon, U. Ordering genes in a flagella pathway by analysis of expression kinetics from living bacteria. *Science* 2001, **292**: 2080–2083.

Mangan, S., Itzkovitz, S., Zaslaver, A., and Alon, U. The incoherent feed-forward loop accelerates the response-time of the gal system of *Escherichia coli*. *Journal of Molecular Biology* 2006, **356**: 1073–1081.

Mangan, S., Zaslaver, A., and Alon, U. The coherent feedforward loop serves as a sign-sensitive delay element in transcription networks. *Journal of Molecular Biology* 2003, **334**(2): 197–204.

Rosenfeld, N., Elowitz, M. B., and Alon, U. Negative autoregulation speeds the response times of transcription networks. *Journal of Molecular Biology* 2002, **323**(5): 785–793.

Shen-Orr, S. S., Milo, R., and Alon, U. Network motifs in the transcriptional regulation network of *Escherichia coli*. *Nature Genetics* 2002, **31**(1): 64–68.

Tabor, J. J., Salis, H. M., Simpson, Z. B., et al. A synthetic genetic edge detection program. *Cell* 2009, **137**: 1272–1281.

Zaslaver, A., Mayo, A. E., Rosenberg, R., Bashkin, P., Sberro, H., Tsalyuk, M., Surette, M. G., and Alon, U. Just-in-time transcription program in metabolic pathways. *Nature Genetics* 2004, **36**(5): 486–491.

PROBLEMS

PROBLEM 7.1
A Coherent Feed-Forward Loop with Repression

In the section entitled "Boolean Analysis of the Most Common Internally Consistent Feed-forward Motif Identified in *E. coli*," we examined the coherent type 1 feed-forward loop, which is the most abundant type of feed-forward loop in biological networks. Here, let's use the notation we developed in that section to look at a different type of coherent feed-forward loop in which p_x^* induces p_y expression, but both p_x^* and p_y^* repress p_z expression. Lumping expression and activation of p_x and p_y into single equations, the Boolean rules that describe this circuit are:

$$\text{Activation\_}p_x = \text{IF } (s_x)$$
$$\text{Activation\_}p_y = \text{IF } (p_x^*) \text{ AND } (s_y)$$
$$\text{Expression\_}p_z = \text{IF NOT } (p_x^*) \text{ AND NOT } (p_y^*)$$

$$p_x^* = \text{IF (Activation\_}p_x) \text{ AFTER SOME TIME}$$
$$p_y^* = \text{IF (Activation\_}p_y) \text{ AFTER SOME TIME}$$
$$p_z = \text{IF (Expression\_}p_z) \text{ AFTER SOME TIME}$$

a. Draw a diagram of the feed-forward loop for this system.

b. Construct a state diagram for this motif for the case in which $s_x = 1$ and $s_y = 1$. Start from $p_x^* = p_y^* = p_z = 0$ and fill in the rows in your matrix until you reach a stable state.

c. What happens when you remove s_x from the system in (b)? Add another column to your state diagram to answer this question.

d. Use your matrix from (b) to graph the time dynamics of p_x^*, p_y^*, and p_z in the feed-forward loop. Begin with the addition of s_x and s_y, graph the progression of the system until it reaches a steady state, then remove s_x and follow the system to the new steady state. Are there delays in the dynamics of p_z?

e. Repeat your state diagram analysis from (b) but change the motif so that the equation for p_z^* contains an OR instead of an AND gate. Are the dynamics of p_z delayed now?

PROBLEM 7.2
A Synthetic Edge-Detection System

Here, we analyze a synthetic transcriptional regulatory circuit designed to detect the edges of an image of light and produce a dark pigment in response (inspired by an actual circuit described by Tabor et al. in 2009). The circuit consists of two signals as inputs: Darkness and Light. There are three proteins (p_x, p_y, and p_z), the first two of which also have active forms. The Boolean rules that describe the system are:

$$\text{Activation\_}p_x^* = \text{IF (Darkness)}$$
$$\text{Activation\_}p_y^* = \text{IF }(p_x^*) \text{ AND (Light)}$$
$$\text{Expression\_}p_z = \text{IF }(p_x^*) \text{ AND }(p_y^*)$$
$$p_x^* = \text{IF (Activation\_}p_x) \text{ AFTER SOME TIME}$$
$$p_y^* = \text{IF (Activation\_}p_y) \text{ AFTER SOME TIME}$$
$$p_z = \text{IF (Expression\_}p_z) \text{ AFTER SOME TIME}$$
$$\text{Pigment} = \text{IF }(p_z)$$

a. Draw a diagram of the feed-forward loop in this system.

b. Draw the state matrix for the system. Note that in this case, the system must have at least one of the signals (do not make a column for $s_x = s_y = 0$). Circle any stable states.

c. For the schematic below, indicate which numbered regions would produce Pigment. Justify your answer.

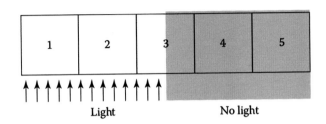

PROBLEM 7.3

Advanced Analysis of an Incoherent Feed-Forward Loop

Let's examine one of the most common regulatory motifs, the incoherent feed-forward loop, shown below. Your goal is to analyze this system, first using ODEs (Chapter 3) and then using a stochastic simulation (Chapter 6). We assume that p_x and p_y are always active (signals s_x and s_y are always present). We also assume that the expression of p_x depends on s_x such that production rate = max production rate · $s_x = \beta_x \cdot s_x$. The Boolean rules for this system are:

$$p_x = \text{IF } (s_x > 0)$$
$$p_y = \text{IF } (p_x > K_{xy})$$
$$p_z = \text{IF } (p_x > K_{xz}) \text{ AND NOT } (p_y > K_{yz})$$

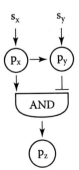

Constants are defined as:

β_x = maximal production rate of p_x

β_y = maximal production rate of p_y

β_z = maximal production rate of p_z

α_x = rate of dilution/degradation of p_x

α_y = rate of dilution/degradation of p_y

α_z = rate of dilution/degradation of p_z.

Assume that you start with no p_x, p_y, or p_z.

a. Write an ODE for the change in $[p_x]$ over time.

b. Write an ODE for the change in $[p_y]$ over time.

c. Assume that s_x is always present at a maximum value of 1, and that the levels of p_x are higher that the threshold for p_y expression ($[p_x] > K_{xy}$). Write the analytical solution for $[p_y](t)$ for the ODE you wrote in (c).

d. Assume that $[p_x] > K_{xy}$ and $[p_x] > K_{xz}$, but initially $[p_y] < K_{yz}$. Write the ODE for $[p_z]$ and solve it analytically for $[p_z](t)$.

e. Continue to assume the same level of p_x ($[p_x] > K_{xy}$ and $[p_x] > K_{xz}$), but the level of p_y is now high enough to stop the expression of p_z ($[p_y] > K_{yz}$). Write an ODE for the levels of p_z over time for this case.

f. Use MATLAB and all of the preceding expressions to plot the levels of p_x, p_y, and p_z over time. Use the following information:

$$s_x = 0 \quad \text{for} \quad -1 \le t < 0$$

$$s_x = 1 \quad \text{for} \quad 0 \le t < 10$$

Constants:

$$\beta_x = 1 \text{ mM/h}$$

$$\beta_y = 1 \text{ mM/h}$$

$$\beta_z = 1 \text{ mM/h}$$

$$\alpha_x = 1/h$$

$$\alpha_y = 1/h$$

$$\alpha_z = 1/h$$

$$K_{xz} = 0.4 \text{ mM}$$

$$K_{xy} = 0.4 \text{ mM}$$

$$K_{yz} = 0.5 \text{ mM}$$

Initially, $[p_x] = [p_y] = [p_z] = 0$ mM.

g. Previously, we assumed that the signal s_x was available for full activation. Now, consider submaximal signaling by s_x. Using the rate constants and activation thresholds in (f), plot the dynamics of $[p_x]$ for $s_x = [0.1:0.1:1]$. What behavior do you see in the response of $[p_z]$ to different levels of s_x? What function might this behavior serve in a living cell?

h. Increase K_{xy} to 0.8 and run your simulation again for $s_x = [0.1:0.1:1]$. How does p_z expression behave in response to various levels of s_x? How might this type of regulation be useful for biological systems?

i. Build and fill a state diagram for the network and circle the stable states. Does this analysis agree with your results from (f)? Why or why not?

j. We will now examine this network using a stochastic simulation. Again, we assume that p_x and p_y will always be active (s_x and s_y are abundant). The reactions are:

$$p_x \text{ expression: } DNA_x \rightarrow DNA_x + p_x$$

$$p_y \text{ expression: } DNA_y + p_x \rightarrow DNA_y + p_x + p_y$$

$$p_z \text{ expression: } DNA_z + p_x \rightarrow DNA_z + p_x + p_z$$

$$p_y\text{:DNA association: } DNA_z + p_y \rightarrow DNA_{z,py}$$

$$p_y: \text{DNA dissociation: } DNA_{z,py} \rightarrow DNA_z + p_y$$

$$p_x \text{ degradation: } p_x \rightarrow 0$$

$$p_y \text{ degradation: } p_y \rightarrow 0$$

$$p_z \text{ degradation: } p_z \rightarrow 0$$

Let the stochastic rate constants be:

$$c_{pxExpression} = 0.8 \text{ proteins/DNA/min}$$

$$c_{pyExpression} = 0.8 \text{ proteins/DNA/min}$$

$$c_{pzExpression} = 0.8 \text{ proteins/DNA/min}$$

$$c_{pyBindingDNAz} = 0.8 \text{ associations/DNA/protein/min}$$

$$c_{pyUnbindingDNAz} = 0.4 \text{ dissociations/complex/min}$$

$$c_{pxDegradation} = 0.001 \text{ degradations/protein/min}$$

$$c_{pyDegradation} = 0.001 \text{ degradations/protein/min}$$

$$c_{pzDegradation} = 0.001 \text{ degradations/protein/min}$$

Assume that you begin with no p_x, p_y, or p_z, but you have one molecule each of DNA_x, DNA_y, and DNA_z. Use MATLAB to write a Gillespie algorithm and run it for 500,000 steps (you may choose to only store every 10th time point). Describe the system's behavior and interpret it in biological terms.

k. What happens if you change the relative protein activities such that p_x promotes the production of p_y with 10-fold less affinity than it does for p_z? (Hint: change a rate constant.) Describe resulting behavior and compare your results qualitatively with your results from (g), (i), and (j).

PROBLEM 7.4
A Single-Input Module

Consider the metabolic pathway in the illustration, which converts glutamate to ornithine in E. coli (slightly modified from Figure 7.15):

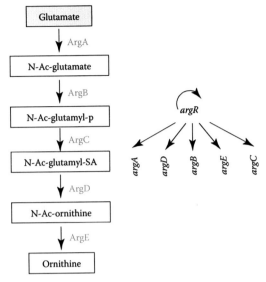

The genes in this pathway are all regulated by the transcription factor ArgR via the single-input motif depicted at right.

a. List two main benefits to the cell of regulating all of these genes with ArgR.

b. Assume that when a signal to activate ArgR is present (s_{ArgR}), the change in ArgR production can be defined as:

$$\frac{d\left[ArgR^*\right]}{dt} = \beta_{ArgR} - \alpha_{ArgR}\left[ArgR^*\right]$$

The change in all of the other gene products can be defined as:

$$\frac{d\left[Arg_i\right]}{dt} = \beta_i\,\theta\left(\left[ArgR^*\right] > K_{ArgR\,Argi}\right) - \alpha_i\left[Arg_i\right]$$

where $i = \{A, B, C, D, E\}$. Given that s_{ArgR} is added at $t = 0$ min and removed at $t = 7$ min, find values for all $K_{ArgR,Argi}$ such that Arg_i will be induced at equal time intervals within the range $[0, t([ArgR] = [ArgR]_{ss}/2)]$, where ss denotes steady state. Note that the gene with the highest threshold parameter will obey $K_{ArgR,Argi} = [ArgR]_{ss}/2$. Use values of 1/min and 1 µM/min for all α and β terms, respectively.

c. Use MATLAB with the information presented to plot [*ArgR**] as well as the expression of all Arg proteins over the time range [0:14] minutes (choose the most appropriate time step). What can you say about the timing of the decay of the various genes, as compared to the timing of induction? Specifically, describe the order and the duration of the decays.

CHAPTER 8

Signal Transduction

LEARNING OBJECTIVES

- Derive the equations for receptor-ligand binding

- Extend the receptor-ligand system to larger complexes

- Adapt the equations to incorporate localization

- Understand how model parameters are determined from raw data

Signal transduction is the mechanism by which cells sense and react to factors in the external environment. To model signal transduction, we must mathematically describe three new aspects of the underlying networks (Figure 8.1): (1) the binding of **ligands** to receptors, (2) the formation of larger protein complexes, and (3) localization within cellular compartments. After covering these three areas, I'll introduce you to one of the first larger-scale models of cellular signaling, which we'll examine in detail.

RECEPTOR-LIGAND BINDING TO FORM A COMPLEX

Let's first look at how the signal is initially detected. A cell might sense the presence of food, an enemy such as a virus or microbe, or other friendly cells in its natural environment. The particular molecules that are detected (for example, a molecule found on the virus surface or a protein that is secreted by a cell) are often called ligands, and bind to proteins called receptors on the cell surface (Figure 8.2). Receptor proteins typically span the cell membrane, so a conformational change on one side of the membrane can alter the protein's binding affinity or kinetic properties on the other side. Active receptor

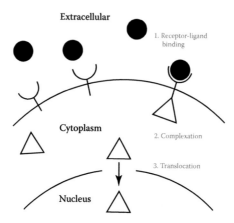

FIGURE 8.1 Schematic of a simplified signal transduction network. We will need to model the three critical aspects of signal transduction highlighted here. Note that this schematic depicts three spatially distinct compartments: the environment external to the cell (the extracellular space), the area inside the cell but outside the cell's nucleus (the cytoplasm), and the nucleus. Noneukaryotic cells, which lack nuclei, nonetheless also contain spatially distinct compartments that should be considered during modeling.

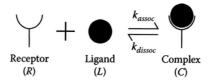

FIGURE 8.2 Schematic of the binding between a receptor and a ligand to form a complex. This binding reaction is reversible and can be described by the parameters k_{assoc} and k_{dissoc}.

proteins then recruit adaptor proteins, which bind the receptor protein domains inside the cell and recruit still other proteins. As the signal is transduced (moved from protein to protein and from one part of the cell to another), it can be amplified (by one protein activating many others) or integrated (as a protein responds to many signals at once). A common final outcome of signaling involves the movement of certain molecules to another cellular compartment; for example, a transcription factor may move from the cytoplasm into the nucleus and induce the transcription of genes.

First, consider the most basic representation of receptor-ligand binding: the association of a single ligand molecule to a single receptor to form a complex, which can also dissociate into its component parts (Figure 8.2).

Note the conceptual similarity to Figure 3.8, the formation of a DNA-protein complex. By analogy to that system, we can write an ODE describing the amount of complex:

$$\frac{dC}{dt} = production - loss$$

$$= k_{assoc}R[L] - k_{dissoc}C$$

(8.1)

where C represents the number of complexes in the cell, R is the number of receptors, and L is the concentration of ligand in the external environment; thus, in our equations, the concentration notation for L is used but not for C or R. Again, we rely on the assumption of mass action kinetics to describe the production and loss of complex. In addition, binding and unbinding reactions such as this reaction (on the order of seconds or even faster) typically occur much faster than protein synthesis or degradation (minutes to hours, as discussed in Section I), so we can assume that the total amounts of receptor and ligand, both bound and unbound, remain constant for this analysis. With this assumption, we can write a conservation relationship, for example by considering the receptors:

$$R_{total} = R + C$$

(8.2)

We can write a similar relationship for the ligand—but not so fast! First, we need to think about units. As I hinted, R and C are generally reported in terms of molecules/cell because they are attached to cells. In contrast, the amount of ligand in the external environment is reported as a concentration, for example, with units of moles/liter (M), because the free ligand is in the environment and independent of the cells. As a result, the units for L and C do not match, and we have to perform a conversion. Avogadro's number (N_{av}) describes the number of molecules/mole ($6.022 \cdot 10^{23}$), so we can convert M to molecules/liter.

To convert from molecules/liter to molecules/cell, we also need to know how many cells are in a given volume, which varies depending on the cell type and the culture conditions. For example, if we grow cells from the common mouse fibroblast line 3T3 in a 10-cm dish, we might expect to have ~1 million cells in 10 mL of culture. If we call the cells/volume ratio n, then the conservation relationship for the ligand is

$$[L]_0 = \frac{n}{N_{av}}C + [L]$$

(8.3)

The total amount of ligand is equal to the initial amount of ligand added to the medium ($[L]_0$). Another consequence of the structure of Equation 8.1 and the difference in units for R, C, and $[L]$ is that the units of k_{dissoc} (1/time) and k_{assoc} (1/concentration/time) are different as well.

If we substitute Equations 8.2 and 8.3 into Equation 8.1, we obtain:

$$\frac{dC}{dt} = k_{assoc}\left(R_{total} - C\right)\left(\left[L\right]_0 - \frac{n}{N_{av}}C\right) - k_{dissoc}C \tag{8.4}$$

The result has only one free variable (C), but the equation is still a bit complicated. We can simplify this expression by assuming that the amount of ligand added to the medium is large compared to the number of receptors in the system, implying that $[L]_0 \gg R_{total} > (n/N_{av}) \cdot C$, so that:

$$\frac{dC}{dt} = k_{assoc}\left(R_{total} - C\right)\left[L\right]_0 - k_{dissoc}C \tag{8.5}$$

Is this assumption always correct? Hardly. In fact, in many cases one is required to titrate the concentrations of ligand to very low levels in order to create a biologically relevant system, as you will see later. To address such studies, we will leave Equation 8.4 intact and use numerical solution methods. However, it is certainly a common experimental design to add ligand in excess, and so Equation 8.5 holds for many cases; these types of experimental or physiological details are worth considering before beginning to model any biological system. Rearranging Equation 8.5 leads to:

$$\frac{dC}{dt} = k_{assoc}\left[L\right]_0 R_{total} - \left(k_{assoc}\left[L\right]_0 + k_{dissoc}\right)C \tag{8.6}$$

I hope the form of Equation 8.6 is stirring sweet nostalgia in your consciousness; if not, notice that $k_{assoc}[L]_0 R_{total}$ is a constant, as is $(k_{assoc}[L]_0 + k_{dissoc})$. Does this remind you of anything? How about the equation for protein expression without autoregulation (Equation 3.23)? In both cases, the form is:

$$\frac{d\left(\text{variable}\right)}{dt} = \text{constant}_1 - \text{constant}_2 \cdot \text{variable} \tag{8.7}$$

The first constant is analogous to basal production, whether of protein or of complex, and the second constant represents the rate of loss via dissociation, dilution, or decay.

You already know the dynamic behavior of complex formation as given by Equation 8.7: exponential increase toward a steady state, as in Figure 3.6.

Even though the signal-transduction system that we are exploring here (Figure 8.1) is different and the timescales are different from our system of protein expression without autoregulation in Chapter 3, the similarity in the underlying equation tells you that the overall dynamics of the response will also be similar. You also know that the steady state will be $constant_1 / constant_2$, or in terms of Equation 8.6:

$$C_{eq} = \frac{k_{assoc} [L]_0 R_{total}}{k_{assoc} [L]_0 + k_{dissoc}} \tag{8.8}$$

The subscript eq stands for "equilibrium." A chemical equilibrium is similar to a steady state for our purposes because in both cases the overall molecule concentrations do not change over time. More technically, a chemical equilibrium differs in that it refers to a single reaction that can occur in the forward and reverse directions and where the *net* chemical reaction rate is zero. Even at equilibrium, association and dissociation occur. The other steady states we have considered involve multiple reactions that produce or consume products; the sum total of these activities yields a stable concentration of the products.

We define the dissociation constant as previously:

$$K \equiv k_{dissoc} \big/ k_{assoc} \tag{8.9}$$

and use K to simplify Equation 8.8, similar to previous chapters:

$$C_{eq} = \frac{[L]_0 R_{total}}{[L]_0 + (k_{dissoc}/k_{assoc})} = \frac{[L]_0 R_{total}}{[L]_0 + K} \tag{8.10}$$

In plain English, if we know the total number of receptors per cell (a number that is specific to the cell line and the environmental conditions), the dissociation constant of the complex, and the amount of ligand that was introduced into the environment, we can predict the number of receptor-ligand complexes per cell at equilibrium. In an *in vitro* system, the amount of ligand is controlled by the experimenter, and the other two parameters must be measured experimentally.

APPLICATION TO REAL RECEPTOR-LIGAND PAIRS

Table 8.1 is a compilation of parameters measured by various groups using different ligands and cell types (see Lauffenburger and Linderman, 1996).

TABLE 8.1 Parameters for Receptor-Ligand Pairs

Receptor	Ligand	Cell type	R_T (#/cell)	k_{assoc} (M⁻¹ min⁻¹)	k_{dissoc} (min⁻¹)	K_D (M)
Chemotactic peptide[1]	Formyl-norleucyl-leucyl-phenylalanine	Rabbit neutrophil	5×10^4	2×10^7	0.4	2×10^{-8}
Interferon[2]	Human interferon α_2a	A549	900	2.2×10^8	0.072	3.3×10^{-10}
Tumor necrosis factor receptor[3]	Tumor necrosis factor α	A549	6.6×10^3	9.6×10^8	0.14	1.5×10^{-10}
Insulin receptor[4]	Insulin	Rat fat cells	1×10^5	9.6×10^6	0.2	2.1×10^{-8}
Epidermal growth factor receptor[5]	Epidermal growth factor	Fetal rat lung	2.5×10^4	1.8×10^8	0.12	6.7×10^{-10}
Integrins[6]	Fibronectin	Fibroblasts	5×10^5	7×10^5	0.6	8.6×10^{-7}

These parameters were previously measured experimentally, and are compiled in Lauffenburger, D. A., Linderman, J. *Receptors: Models for Binding, Trafficking, and Signaling.* New York: Oxford University Press, 1996. Modified with permission of Oxford University Press, USA. Original references appear as superscripted numerals as follows: [1]Ciechanover, A., Schwartz, A. L., Dautry-Varsat, A., and Lodish, H. F. Kinetics of internalization and recycling of transferrin and the transferrin receptor in a human hepatoma cell line. Effect of lysosomotropic agents. *The Journal of Biological Chemistry* 1983, **258**(16): 9681–9689. [2]Mellman, I. S. and Unkeless, J. C. Purification of a functional mouse Fc receptor through the use of a monoclonal antibody. *The Journal of Experimental Medicine* 1980, **152**(4): 1048–1069. [3]Bajzer, Z., Myers, A. C., and Vuk-Pavlovic, S. Binding, internalization, and intracellular processing of proteins interacting with recycling receptors. A kinetic analysis. *The Journal of Biological Chemistry* 1989, **264**(23): 13623–13631. [4]Hughes, R. J., Boyle, M. R., Brown, R. D., Taylor, P., and Insel, P. A. Characterization of coexisting alpha 1- and beta 2-adrenergic receptors on a cloned muscle cell line, BC3H-1. *Molecular Pharmacology* 1982, **22**(2): 258–266. [5]Waters, C. M., Oberg, K. C., Carpenter, G., and Overholser, K. A. Rate constants for binding, dissociation, and internalization of EGF: effect of receptor occupancy and ligand concentration. *Biochemistry* 1990, **29**(14): 3563–3569. [6]Akiyama, S. K. and Yamada, K. M. The interaction of plasma fibronectin with fibroblastic cells in suspension. *The Journal of Biological Chemistry* 1985, **260**(7): 4492–4500.

In Table 8.1, you can see that the number of receptors reported by these labs ranged from about 1,000 to tens of thousands of receptors per cell, providing you with a ballpark estimate. If you compare the values of k_{assoc} and k_{dissoc}, you will see that the former are much greater than the latter. The result is a value of K that is very small, meaning that these particular ligands and receptors bind tightly; the forward reaction is much more likely to occur than the reverse reaction.

Let's zero in on one of these proteins in detail. My favorite protein in Table 8.1 is tumor necrosis factor alpha (TNF-α). TNF-α is a cytokine, a protein that is used for intercellular signaling, and it plays an important role in the immune system. The receptor protein for TNF-α was creatively named <u>t</u>umor <u>n</u>ecrosis <u>f</u>actor <u>r</u>eceptor, or TNFR. The interaction between TNF-α and TNFR was studied in A549 cells, which are derived from the lung.

Given the values of R_{total} and K_D from Table 8.1, and using Equation 8.10, we can plot the number of complexes per cell at equilibrium for every starting concentration of ligand (Figure 8.3). This plot is particularly interesting to me because I stimulate cells with TNF-α all the time, and the ligand concentration that I choose turns out to have a critical impact on cell behavior. Like those in many other labs in my field, I often use a ligand concentration of 10 ng/mL, an amount that ensures that excess ligand is present relative to receptor without causing the cells to die. Figure 8.3

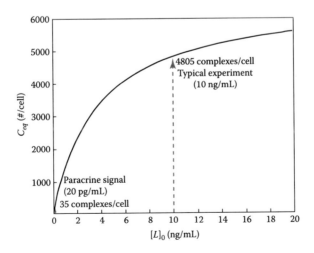

FIGURE 8.3 Binding of a ligand to a receptor, with values taken from the TNF data in Table 8.1 and using Equation 8.10. Dashed red lines highlight the two ligand concentrations discussed in the text.

indicates that you would expect ~5,000 receptors per cell to be bound in the presence of 10 ng/mL TNF-α, meaning that ~75% of the total number of receptors are bound.

To determine the effect of lower ligand concentrations, we recently started performing experiments with 500-fold less TNF-α (20 pg/mL); at that concentration, 35 molecules per cell are bound to receptors (Figure 8.3). Of course, the number 35 is an average, so some cells may bind ~100 ligands, and others may not bind any at all. In our experiments, we see that some cells become activated at those low concentrations, but other cells do not. The number of ligands is low enough that stochasticity can play a significant role in the dynamics (as discussed in Problem 8.2, which addresses stochasticity in receptor-ligand binding).

Low concentrations of ligand pose further problems to our analysis. Recall that to simplify Equation 8.4 into Equation 8.5, we assumed that the ligand concentration was large with respect to the number of available receptors. You can see from Figure 8.3 that this assumption is not valid at 20 pg/mL of ligand, so we need to use Equation 8.4 to plot the dynamics of complexation. However, you can plot the dynamics of Equation 8.4 numerically. This is no problem for us thanks to our training in Section I! But you'll need one more parameter: the number of cells per volume. As mentioned previously, for the commonly used 3T3 cell line, we assume ~1 million cells per 10 mL of culture.

The plots for both 10 ng/mL and 20 pg/mL of TNF-α appear in Figure 8.4. The higher concentration of ligand induces a rapid rise to a large

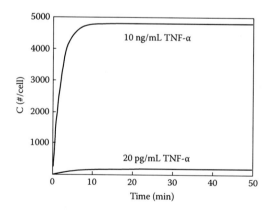

FIGURE 8.4 Dynamics of complex formation for initial ligand concentrations of 10 ng/mL and 20 pg/mL TNF. The solution to Equation 8.4 was determined numerically.

steady-state number of complexes per cell; the smaller concentration leads to a slower response and many fewer complexes per cell.

FORMATION OF LARGER COMPLEXES

Now that we've covered binding of a ligand to its receptor, we can start to consider further interactions. For example, once TNF-α binds to TNFR, the receptors bind together to form units of three, or trimers, which can then bind to other proteins inside the cell and form even larger complexes. Let's see how we can incorporate these complex-building interactions (step 2 in Figure 8.1) into our models. The simplest addition is to add a single protein to a ligand-receptor complex, generating a ternary complex. Figure 8.5 shows our new network, which includes ligand-receptor binding but has a third component, an adaptor protein, that binds the cytoplasmic side of the receptor—regardless of whether or not the ligand is bound. Each association and dissociation reaction has its own rate constant.

PRACTICE PROBLEM 8.1
Write the ODEs describing the changes in C_1, C_2, and C_3, as well as any conservation relationships depicted in Figure 8.5.

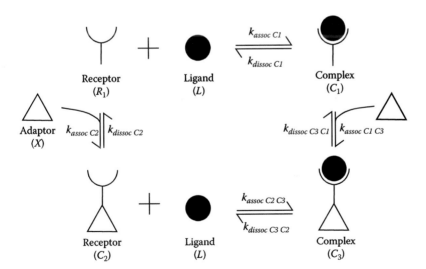

FIGURE 8.5 Schematic of our new network that includes a ternary complex. The adaptor protein binds the receptor to further transmit the signal introduced by binding of the ligand to the receptor.

SOLUTION

The equations are shown below. As you can see, adding one dimension (an adaptor protein) adds a significant amount of complexity; not only has the number of equations changed, but also the number of interactions encoded in each equation. This complexity is a major challenge of making large, detailed models. Here, $[L]_0$ is the total concentration of ligand (in all bound and free forms), R_T is the total amount of receptor (in all bound and free forms), and X_T is the total amount of adaptor (in all bound and free forms).

$$\frac{dC_1}{dt} = k_{assocC1} \cdot R_1 \cdot [L] - k_{dissocC1} \cdot C_1 + k_{dissocC3C1} \cdot C_3 - k_{assocC1C3} \cdot C_1 \cdot X$$

$$\frac{dC_2}{dt} = k_{assocC2} \cdot R_1 \cdot X - k_{dissocC2} \cdot C_2 + k_{dissocC3C2} \cdot C_3 - k_{assocC2C3} \cdot C_2 \cdot [L]$$

$$\frac{dC_3}{dt} = k_{assocC1C3} \cdot C_1 \cdot X - k_{dissocC3C1} \cdot C_3 - k_{dissocC3C2} \cdot C_3 + k_{assocC2C3} \cdot C_2 \cdot [L]$$

$$R_T = R_1 + C_1 + C_2 + C_3$$

$$X_T = X + C_2 + C_3$$

$$[L]_0 = [L] + \frac{n}{N_{av}} C_1 + \frac{n}{N_{av}} C_3$$

PROTEIN LOCALIZATION

Now it's time to consider localization, the third molecular activity highlighted in Figure 8.1. The key issue here is that molecules can move to certain areas in cells, and this motion can affect signal transduction and therefore cellular behavior. There are many examples of this movement in **eukaryotic** systems; for example, transcription factors like NF-κB (see below) are generally activated by an interaction in the cytoplasm and then translocate to the nucleus to bind DNA.

Let's consider the most basic way to represent protein localization mathematically: the compartment method. First, we define the compartments of interest. Some common examples are the nucleus, the cytoplasm, the mitochondria, and the environment external to the cell. Any area that is separated from other areas by a membrane is worth considering if the proteins of interest appear there. In some cases, a protein translocates to a cellular membrane, so that the membrane would also be a compartment.

A good knowledge of the biological system you are modeling will help you to define these compartments.

Next, we write the system of ODEs. We create a variable for every molecule that we are tracking, as before, but now we also create extra variables for each molecule to represent each compartment where the molecule could be. In the case of the eukaryotic transcription factor, instead of having a single concentration term [*Protein*], we would have two: [*Protein*$_{cytoplasm}$] and [*Protein*$_{nucleus}$]. We then include extra terms in the ODEs to reflect movement between the compartments, so that we define not only *production* and *loss* terms, but also *transport in* and *transport out* terms.

We can use endocytosis, the process of bringing material into the cell via membrane invagination to form endosomes (Figure 8.6), as an example of how to include localization in a model. As previously, we depict the interaction of a receptor and a ligand to form a complex. Both free receptors and complexes can be endocytosed into endosomes. Some of

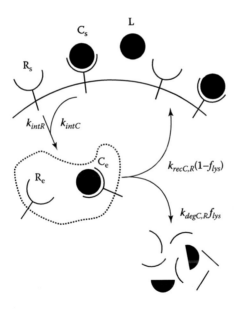

FIGURE 8.6 Schematic of the receptor-ligand complexation system, including compartmentalization. The receptor on the surface R_s binds the free ligand in the environment L to generate a receptor-ligand complex on the surface C_s. This complex is internalized into the cell with rate constant k_{intC}, as is the free receptor with rate constant k_{intR}; both molecules are sequestered in an endosome (C_e and R_e, respectively), which is depicted with a dotted line. A fraction of the molecule pool f_{lys} is degraded in the cell with rate constants k_{degC} and k_{degR}; the remaining fraction $(1 - f_{lys})$ is recycled with rate constants k_{recC} and k_{recR}.

the endosomes are eventually routed to lysosomes, where their contents (ligands, receptors, and complexes) are degraded; other endosomes return to and fuse with the cell membrane to release and recycle their occupants.

Let's write the ODEs. As before, we assume a large concentration of ligand so that $[L] \approx [L]_0$, and no further ODE is needed to describe $[L]$ (see Equation 8.5 and corresponding discussion). For our purposes, the receptor and complex exist in two compartments (we will not explicitly include lysosomes here; notice that there is no lysosome included in the degradation process in Figure 8.6): the cell surface (denoted with an s subscript) and the endosome (e subscript). As a result, we will need four ODEs to include both the receptor and complex in each compartment.

$$\frac{dR_s}{dt} = \underbrace{-k_{assoc}[L]_0 R_s + k_{dissoc}C_s}_{association/dissociation} - \underbrace{k_{intR}R_s}_{internalization} + \underbrace{k_{recR}\left(1-f_{lys}\right)R_e}_{recycling} \quad (8.11)$$

$$\frac{dC_s}{dt} = \underbrace{k_{assoc}[L]_0 R_s - k_{dissoc}C_s}_{association/dissociation} - \underbrace{k_{intC}C_s}_{internalization} + \underbrace{k_{recC}\left(1-f_{lys}\right)C_e}_{recycling} \quad (8.12)$$

$$\frac{dR_e}{dt} = \underbrace{k_{intR}R_s}_{internalization} - \underbrace{k_{recR}\left(1-f_{lys}\right)R_e}_{recycling} - \underbrace{k_{degR}f_{lys}R_e}_{lysosomal\ degradation} \quad (8.13)$$

$$\frac{dC_e}{dt} = \underbrace{k_{intC}C_s}_{internalization} - \underbrace{k_{recC}\left(1-f_{lys}\right)C_e}_{recycling} - \underbrace{k_{degC}f_{lys}C_e}_{lysosomal\ degradation} \quad (8.14)$$

There is also a new conservation relationship: $R_{tot} = R_s + R_e + C_s + C_e$.

The terms in Equations 8.11–8.14 are labeled (underlined and categorized) so that we can focus on similar terms together. First, notice that the association/dissociation terms are identical to what we have already studied in this chapter. However, these reactions are now limited to a particular compartment, the cell surface. Next, there are internalization terms for the receptor and complex, which are also expressed in terms of mass action kinetics: The rate of internalization is proportional to the concentration of receptor or complex on the surface. Once inside an endosome, a fraction f_{lys} of the receptors and complexes is degraded in the cell with rate constants k_{degR} and k_{degC}, under the assumption of mass action kinetics. Finally, the fraction of receptors and complexes that are recycled is $(1-f_{lys})$, and recycling occurs with rate constants k_{recR} and k_{recC}.

THE NF-κB SIGNALING NETWORK

Now that we have used some simple examples to explore a few fundamental considerations in modeling cell signaling, let's look into a significantly more complicated signaling model. As they say in the news, "If it bleeds, it leads," so I'm leading with a corpse in Figure 8.7—a corpse that provides a major clue to how animals fight infection. A dead fly in Jules Hoffmann's lab at the Centre National de la Recherche Scientifique was found covered with a pathogenic fungus (that stringy-looking stuff indicated in red, for those of you who aren't accustomed to fruit fly forensics). Normally, flies are somewhat resistant to this fungus, but this fly harbors a mutation in a single gene that makes it much more susceptible to infection.

The gene was named *Toll*, which means "cool!" in German. Why was the discovery of this gene so cool? It provided a major insight into the molecular mechanisms of the innate immune response, mechanisms that are conserved in humans. As a result of this discovery, Hoffmann won the Nobel Prize in 2011. Now that's what I call *toll*!

It turns out that Toll is a receptor protein, and humans have many variants of this protein, which are called "Toll-like receptors" or TLRs. These receptors specifically bind to general classes of molecules that would be associated with an infection. As you can see in Figure 8.8, TLR7 and TLR8 bind viral single-stranded RNA; TLR4 binds bacterial

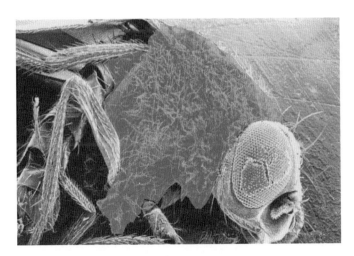

FIGURE 8.7 *Drosophila* infected with a fungus due to its mutation in *Toll*. The fungus is pseudocolored in red. (Reprinted from Lemaitre, B., Nicolas, E., Michaut, L., Reichhart, J. M., and Hoffmann, J. A. *Cell* 1996, **86**(6): 973–983, with permission from Elsevier.)

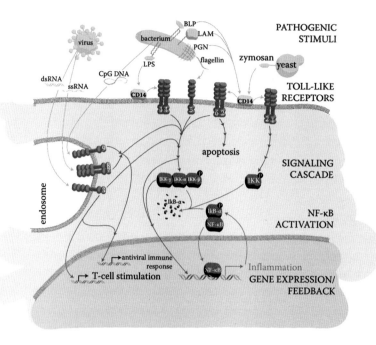

FIGURE 8.8 The Toll-like signaling pathways mediate the cellular response to infection. The TLRs can sense ligands, including molecules found on or inside viruses, bacteria, and yeast; each TLR binds to specific types of molecules (for example, TLR4 binds specifically to lipopolysaccharide, a molecule specific to bacteria). Some of the TLRs occur directly on the cell membrane; others are located in endosomes, where they access cellular debris from pathogens (such as the unusual nucleotides that make up some viral genomes). Once a TLR is bound, it undergoes a conformational change that is sensed by proteins inside the cell, eventually leading to the activation of IKK. IKK phosphorylates the IκB proteins, which in turn are degraded, allowing NF-κB to enter the nucleus and regulate the expression of hundreds of genes, including at least one of the IκBs.

membrane-associated molecules called lipopolysaccharides. The stimulation of different receptors can lead to different responses.

However, one part of the response appears to be consistent across TLRs, and we will call this part the nuclear factor kappa B (NF-κB) circuit. NF-κB is the name of a family of transcription factors. In unstimulated cells, NF-κB resides in the cytoplasm and is bound to one molecule of a family of inhibitory proteins called inhibitor of kappa B (IκB). These IκBs come in at least three different varieties—IκBα, IκBβ, and IκBγ—and prevent NF-κB from entering the nucleus by covering up its **nuclear localization signal**, a part of the protein that allows it to be carried across the nuclear envelope. However, when the TLRs bind to a ligand, a cascade of signaling occurs that leads to activation of a protein called inhibitor of kappa B kinase (IKK). IKK phosphorylates the IκB proteins, labeling them for degradation by the cell. The removal of IκB exposes the nuclear localization signal, enabling NF-κB to translocate to the nucleus.

Once in the nucleus, NF-κB initiates the transcription of hundreds of genes, which in turn can dramatically affect cellular behavior. NF-κB is sometimes described as being at the "crossroads of life and death" because its activation can lead to programmed cell death (**apoptosis**) or more rapid cellular division (proliferation) in response to the source and context of an infection. Moreover, at least one of the IκB proteins, IκB-α, is also induced as a result of the activity of NF-κB. Increased IκB-α expression results in negative feedback: NF-κB is inactivated by binding of the new IκB until this new IκB has also been labeled by IKK and subsequently degraded, resulting in NF-κB re-activation.

The cycle of NF-κB activation and translocation leading to IκB-α expression and subsequent binding to NF-κB, followed by IκB labeling and degradation to reactivate NF-κB, leads to shuttling of NF-κB back and forth between the nucleus and cytoplasm. Figure 8.9 contains images of NF-κB oscillations, as well as time-course plots of the amount of nuclear NF-κB increasing and decreasing at regular intervals.

A DETAILED MODEL OF NF-κB ACTIVITY

These interesting dynamics attracted the interest of the modeling community just over a decade ago. Most notably, Alex Hoffmann, a biochemist (now at the University of California at San Diego), and Andre Levchenko, a physicist (now at Johns Hopkins University), got together with others at Caltech to build a mathematical model of the key aspects of NF-κB dynamics. On a personal note, Hoffmann and Levchenko's work had a huge impact on me as a graduate student; I changed my life plan completely after graduate school to start working in this field! At the time, it was the one of the best-characterized models of mammalian signaling, and the combination of modeling and experimentation was powerful, as you will see.

The Hoffmann–Levchenko model includes 25 ODEs that are much more complex than anything treated in this book so far, so we will use a few figures to keep everything straight. We start with the main feedback circuit in the model (Figure 8.10), which encompasses the binding and unbinding of NF-κB, IκB, and IKK. I hope that Figure 8.10 reminds you of the ternary

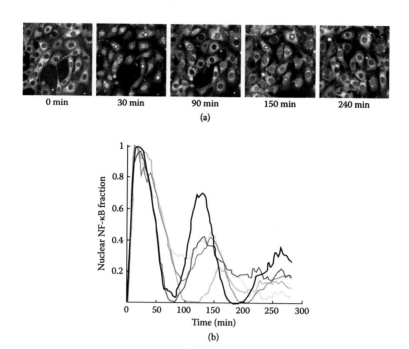

FIGURE 8.9   NF-κB shuttles between the nucleus and the cytoplasm when stimulated by the ligand TNF-α. (a) NF-κB movement into and out of nuclei (outlined in red) in single cells over time. NF-κB was labeled with a fluorescent protein and appears as bright spots within the outlines in these images. (b) Quantitation of the amount of NF-κB within the nuclear outlines in individual cells. The regular cycle of increasing and decreasing nuclear NF-κB fraction is often termed an "oscillation." (From Lee, T. K., Denny, E. M., Sanghvi, J. C., Gaston, J. E., Maynard, N. D., Hughey, J. J., and Covert, M. W. *Science Signaling* 2009, 2(93): ra65. Reprinted with permission from AAAS.)

complexation system in Figure 8.5 because the structure and the equations are essentially the same. The one difference is that there are three types of IκB isoforms, so there are three times the number of reactions that can occur (and, consequently, three times the number of parameters).

Next, we have to consider the transcriptional part of the network (Figure 8.11). Transcription of the three IκB genes is modeled, and each gene has its own parameter for basal expression. NF-κB only affects IκB-α in the model; for this effect to occur, NF-κB must be in the nucleus. Hence, the model has one ODE for the change in concentration of cytoplasmic NF-κB over time and another for nuclear NF-κB concentration. The model includes two parameters for translocation to and from the nucleus.

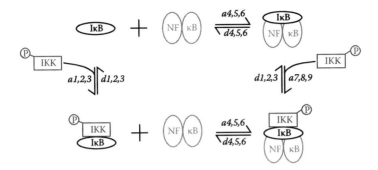

FIGURE 8.10 The binding and unbinding of NF-κB, IκB, and IKK as represented in the Hoffmann–Levchenko model. Parameter names appear near the corresponding reaction arrows. Any of the three IκB isoforms can bind to NF-κB or IKK, which is why three reactions/parameters are denoted by commas. In the parameter names from the original publications, the first number in the subscript corresponds to IκB-α (for example, $a1$ or $d4$), the next to IκB-β ($a2$ or $d5$), and so on. Only active IKK appears in this model.

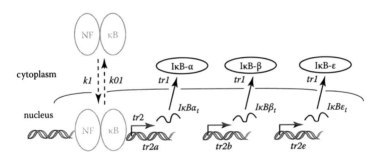

FIGURE 8.11 Transcription and translation of the IκB genes in the model. NF-κB translocates from the cytoplasm to the nucleus and vice versa. Nuclear NF-κB binds the IκB-α promoter, inducing expression of this mRNA transcript ($IκBα_t$). The transcript is then translated, resulting in IκB-α protein in the cytoplasm. IκB-β and IκB-ε are also transcribed and translated, but not under NF-κB control.

IκB mRNA transcripts are only found in the nucleus, and translation leads to new copies of IκB in the cytoplasm. Note that only a single parameter, trl, represents the translation of all three IκBs.

NF-κB is not the only protein or complex that translocates between the nucleus and cytoplasm; free IκBs can also shuttle between compartments (Figure 8.12). Moreover, NF-κB and the IκBs can associate in the nucleus, and the resulting IκB:NF-κB complex can translocate from the nucleus to the cytoplasm.

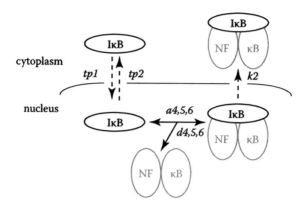

FIGURE 8.12 Translocation of free IκB and IκB complexed with NF-κB, as represented in the model. Note that translocation of the IκB:NF-κB complex is unidirectional. Association and dissociation parameter names are defined as in Figure 8.10.

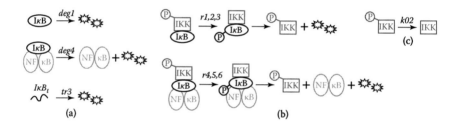

FIGURE 8.13 Degradation in the NF-κB signaling model. (a) Spontaneous degradation. (b) IKK-induced degradation. (c) Adaptation of IKK to the signal.

Degradation plays a key role in this model: Free IκB proteins and IκB:NF-κB complexes both degrade spontaneously, as do the IκB mRNA transcripts (Figure 8.13). Moreover, complexation with IKK increases the degradation rate for free IκB proteins and IκB:NF-κB complexes. Finally, the IKK signal itself decreases over time, a phenomenon called **adaptation**.

Now, we've covered essentially all of the features of the model, so let's write some equations!

PRACTICE PROBLEM 8.2

Write an ODE describing the NF-κB concentration in the cytoplasm. Assume that all reaction rates follow mass action kinetics.

SOLUTION

This equation appears in the original manuscript by Hoffmann et al. (2002). It looks daunting, but don't worry—you already have the tools to analyze it! Just take a deep breath, say "production and loss" three times, and dive in:

$$\frac{d[NFkB]}{dt} = \underbrace{-a4\cdot[IkBa]\cdot[NFkB]-a4\cdot[IKK\_IkBa]\cdot[NFkB]-a5\cdot[IkBb]\cdot[NFkB]}_{association}$$

$$\underbrace{-a5\cdot[IKK\_IkBb]\cdot[NFkB]-a6\cdot[IkBe]\cdot[NFkB]-a6\cdot[IKK\_IkBe]\cdot[NFkB]}_{association}$$

$$\underbrace{+d4\cdot[IkBa\_NFkB]+d4\cdot[IKK\_IkBa\_NFkB]+d5\cdot[IkBb\_NFkB]}_{dissociation}$$

$$\underbrace{+d5\cdot[IKK\_IkBb\_NFkB]+d6\cdot[IkBe\_NFkB]+d6\cdot[IKK\_IkBe\_NFkB]}_{dissociation}$$

$$\underbrace{+r4\cdot[IKK\_IkBa\_NFkB]+r5\cdot[IKK\_IkBb\_NFkB]+r6\cdot[IKK\_IkBe\_NFkB]}_{IKK\ reaction\ degradation}$$

$$\underbrace{+deg4\cdot[IkBa\_NFkB]+deg4\cdot[IkBb\_NFkB]+deg4\cdot[IkBe\_NFkB]}_{spontaneous\ degradation}$$

$$\underbrace{-k1\cdot[NFkB]+k01\cdot[NFkBn]}_{nuclear\ translocation}$$

The loss terms in the equation come from protein-protein association. Cytoplasmic NF-κB can bind any of the IκB isoforms, as well as any isoform bound to IKK. Any of the resulting complexes can also dissociate, leading to production of NF-κB. Free NF-κB can also be produced from a complex as a result of IκB degradation, which can occur spontaneously or be catalyzed by IKK. Finally, the NF-κB can be transported to or from the nucleus. The resulting equation is lengthy, but it is hardly more intellectually challenging than the other equations you have addressed.

PRACTICE PROBLEM 8.3

Write an ODE to describe the concentration of (a) IκB-β and (b) IκB-α mRNA in the nucleus. For (b), you should know that NF-κB is a dimer, and that both subunits bind separate binding sites on the DNA.

SOLUTION TO PART A

The equation for IκB-β is just like the constitutively expressed gene without feedback, as discussed in Chapter 3:

$$\frac{d[IkBb_t]}{dt} = tr2b - tr3 \cdot [IkBb_t]$$

where $tr2b$ is the constant rate of expression, and $tr3$ is the decay rate constant.

SOLUTION TO PART B

The equation for IκB-α is more complicated. Based on what you have already learned, I would expect you to write something like:

$$\frac{d[IkBa_t]}{dt} = tr2a + \frac{k_{trs,max} \cdot [NFkBn]^2}{K_{diss}^2 + [NFkBn]^2} - tr3 \cdot [IkBa_t]$$

Again, there's a basal constant rate of expression $tr2a$ and a decay rate constant $tr3$. However, in this case, there is also an inducible expression rate. $k_{trs,max}$ is the maximum amount of induced expression, and K_{diss} is the binding constant ($k_{dissociation}/k_{association}$) for NF-κB to DNA. The second power comes from the fact that NF-κB binds two binding sites, resulting in cooperativity (see Problem 3.4).

ALTERNATIVE REPRESENTATIONS FOR THE SAME PROCESS

I need to emphasize a critical point to remember when working with real-life models: building them is an art! There are many decisions to be made, not only with regard to which modeling approach to use (I hope you know a lot about that now), but also in terms of how to characterize and parameterize the model. This chapter and the two that follow all contain several "judgment calls." You will likely ask yourself, how did the modelers know to do it that way? The answer is that they didn't know; they most likely just tried some things and kept the one that seemed to match both the experimental data and their intuition most clearly.

For example, here is the equation that Hoffmann and Levchenko et al. used in their model:

$$\frac{d[IkBa_t]}{dt} = tr2a + tr2 \cdot [NFkBn]^2 - tr3 \cdot [IkBa_t] \tag{8.15}$$

Notice that the inducible term is very different from the one we formulated in Practice Problem 8.3. Instead of deriving a model of protein-DNA binding, Hoffmann and Levchenko et al. simply stated that the rate was proportional to the square of the NF-κB concentration. Now, we have two different ODEs describing the concentration of IκB-α mRNA.

How do these two equations compare with each other? We answer this question by plotting the mRNA production (the first two terms on the right side of the equation) as a function of the nuclear NF-κB concentration for both equations. Figure 8.14 contains these plots, using the Hoffmann–Levchenko parameter values for $tr2a$ and $tr2$, and values for K_{diss} and $k_{trs, max}$ that were selected to facilitate comparison. One equation yields a sigmoid curve; the other is parabolic. You can also see that for a certain range of [$NFkBn$], the two equations produce strikingly similar results. The most notable difference is that the sigmoid curve eventually reaches a maximum ($tr2a + k_{trs, max}$), and the parabolic curve increases infinitely. In other words, as long as the simulated values for [$NFkBn$] remain within a certain range, the choice of terms may not have a dramatic impact on the predictions from the model. Outside this range, there will likely be a big difference.

The model output can therefore be sensitive to its underlying structure (for example, whether we chose the Hoffmann–Levchenko term or

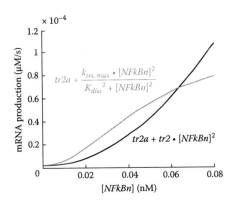

FIGURE 8.14 Simulations of two possible equations for the production of IκB-α mRNA, calculated from the nuclear concentration of NF-κB, as indicated by the equations shown. The trace shown in gray is what you should have derived based on what we've already learned; the darker trace is what Hoffmann and Levchenko actually used. Notice that for a particular range, the calculated mRNA production is roughly the same for either method; beyond ~0.07 nM of nuclear NF-κB, the darker trace rises dramatically, while the gray trace approaches an asymptotic limit.

our own induction term), and its sensitivity can vary depending on the conditions; for example, the model is not so sensitive at low [*NFkBn*] but is more sensitive when [*NFkBn*] becomes large. The output can also be sensitive to the values of particular parameters. For example, varying a single parameter can change the number of fixed points, as we saw in Chapter 4.

SPECIFYING PARAMETER VALUES FROM DATA

The Hoffmann–Levchenko model includes 35 constants, and to produce an output, we need to have a value for each parameter. We can roughly divide the parameters into three sets: (1) parameters that are specified in the literature, (2) parameters that are bounded by the literature, and (3) unknown parameters. Let's examine each set and consider an example for each.

Twenty-three of the parameters in this model were previously specified in the literature when the model was originally constructed. This type of parameter is great for modeling because all you have to do is find the value in the literature and "plug it in." Relatively little guesswork is required. The specified parameters include protein-protein association and dissociation constants (15 parameters total), rate constants for the reaction in which IKK phosphorylates IκB molecules (free or bound to NF-κB; six parameters), as well as the rate constants for degradation of free or bound IκB (two parameters).

Short of becoming a full-fledged experimentalist, as a modeler you can at least start to learn the biology the way it was originally learned: by looking at raw data. To consider the parameters in the NF-κB model, let's examine the original data so that you can appreciate where they came from and what assumptions went into interpreting them.

For example, take a look at the data from which the parameters for IκB-α degradation were inferred. The data were collected by allowing cells to take up an amino acid (methionine) that carries a radioactive isomer of sulfur. The radioactive isomer was incorporated into proteins, including IκB-α, which were then isolated using a specific antibody (immunoprecipitation). The radiolabeled IκB-α was then purified by gel electrophoresis, after which it was detected on a phosphorimager screen.

In this case, the scientists were interested in the half-life of free and IκB-α:NF-κB. They performed a **pulse-chase labeling** experiment by allowing the cells to take up radiolabeled amino acid for several hours (the "pulse"), but then "chasing" the pulse by bathing the cells in radioactive isomer-free medium for 60, 120, or 240 min before immunoprecipitating the IκB-α. Some of the IκB that is labeled by the pulse is degraded during the chase, so the decrease in the amount of labeled IκB is an indicator of the protein's half-life.

Figure 8.15 shows how the raw data are translated into parameters. The radioactive bands (Figure 8.15a) are quantified (Figure 8.15b) using image analysis software, which normally comes with the equipment; you can also use MATLAB's image analysis toolbox for more control. Notice that the *y* axis is labeled "% relative intensity." We do not know exactly how much IκB is in the sample, but we can determine how much more or less IκB is in one sample compared to another sample. Here, the 100% intensity value comes from the darkest band, and the 0% value presumably comes from a control lane with no sample added.

Finally, we need to calculate the value of the half-life based on these quantitative data (Figure 8.15c). We use the simplest and most commonly used means of **parameter estimation**: linear least-squares regression.

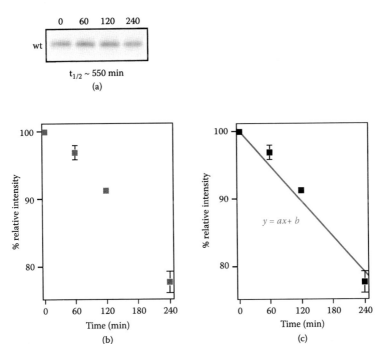

FIGURE 8.15   Pulse-chase labeling followed by immunoprecipitation to determine the half-lives of free IκB-α and NF-κB-bound IκB-α. (a) The raw data are collected from the image; the darker the band on the gel, the more labeled IκB-α is present in the sample. wt, wild type. (b) The gel images are computationally transformed into quantitative data. (c) The protein half-life is determined by fitting a line to the data. (Modified from Pando, M. P., and Verma, I. M. *The Journal of Biological Chemistry* 2000, **275**(28): 21278–21286. © The American Society for Biochemistry and Molecular Biology.)

First, we assume that the data can be well described by a line of the form $y = ax + b$, or:

$$\text{intensity} = (\text{change in intensity over time}) \cdot \text{time} + (\text{initial intensity}) \quad (8.16)$$

Next, we consider each data point in turn and determine its distance from the line in the y direction. This value is called the **residual** and is calculated as:

$$d_i = y_i - (ax_i + b) \quad (8.17)$$

where d_i is the residual for the ith data point. Because d_i can be positive or negative, our goal is to minimize the sum of the squared distances (absolute distances would also be fine) between the data and the line, meaning:

$$\sum_{i=1}^{n} d_i^2 \quad (8.18)$$

or equivalently:

$$\sum_{i=1}^{n} (y_i - ax_i - b)^2 \quad (8.19)$$

where n is the number of data points. Taking the derivative of Equation 8.19, first in terms of a and then of b, yields two ODEs; setting the derivative term in each equation equal to zero will identify the minimum, with two algebraic equations that can be solved simultaneously to obtain a and b.

PRACTICE PROBLEM 8.4
Let's practice linear fitting to obtain parameters! Given the data in Figure 8.15 and Table 8.2, calculate the slope and the intercept of the line that best fits the data.

TABLE 8.2   Data for Practice Problem 8.4

Time (min)	Relative Intensity (%)
0	100
60	97
120	91
240	78

SOLUTION

We can expand Equation 8.19 as:

$$\sum_{i=1}^{n} d_i^2 = (100\% - a \cdot 0 - b)^2 + (97\% - a \cdot 60 - b)^2$$
$$+ (91\% - a \cdot 120 - b)^2 + (78\% - a \cdot 240 - b)^2$$

Taking the derivative with respect to a and b and setting the lefthand side equal to zero yields two equations:

$$0 = 2(0) + 2(97\% - a \cdot 60 - b)(-60) + 2(91\% - a \cdot 120 - b)(-120)$$
$$+ 2(78\% - a \cdot 240 - b)(-240)$$
$$0 = 2(100\% - a \cdot 0 - b)(-1) + 2(97\% - a \cdot 60 - b)(-1)$$
$$+ 2(91\% - a \cdot 120 - b)(-1) + 2(78\% - a \cdot 240 - b)(-1)$$

Simplifying, we obtain

$$b = \frac{183}{2} - 105 \cdot a$$

$$b = \frac{591}{7} - \frac{1260}{7} \cdot a$$

By substitution, we can determine that $a = -0.0943$ and $b = 101.4$. There are also MATLAB (`polyfit`) and Microsoft Excel (`slope, intercept`) functions that can perform this calculation, but it's good to know that you could calculate the solutions yourself if necessary.

Once the line has been specified, we can use it to determine the half-life of the IκB-α bound to NF-κB. You can see from Figure 8.15 that a and b in our system are defined as:

$$\text{intensity} = (\text{change in intensity over time}) \cdot \text{time}$$
$$+ (\text{intensity at time} = 0) \tag{8.20}$$

The half-life is then calculated as:

$$t_{1/2} = -(\text{initial intensity} - \frac{1}{2} \cdot \text{initial intensity})$$
$$/(\text{change of intensity over time}) \tag{8.21}$$

or:

$$t_{1/2} = -\text{initial intensity}/(2 \cdot \text{change of intensity over time}) \quad (8.22)$$

Substituting our a and b,

$$t_{1/2} = -(101.4)/(2 \cdot -0.0943) \text{ min} \quad (8.23)$$

$$= 540 \text{ min}$$

For Figure 8.15, this calculation returns a half-life of ~550 min for IκB-α bound to NF-κB—which is what was reported and used in the model.

The same method can be used to determine the half-life for IκB-α unable to bind NF-κB. Figure 8.16 shows the raw data from an experiment in which IκB-α can't be bound due to a mutation. The free IκB-α has a shorter half-life of ~110 min.

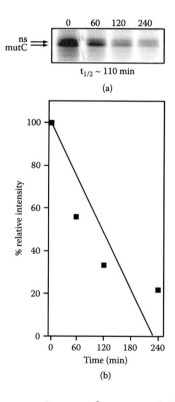

FIGURE 8.16 Pulse-chase experiment with a mutant IκB-α that cannot associate with NF-κB (mutC). ns, nonspecific binding. (Modified from Pando, M. P., and Verma, I. M. *Journal of Biological Chemistry* 2000, **275**(28): 21278–21286. © The American Society for Biochemistry and Molecular Biology.)

Let's take another critical look at Figures 8.15 and 8.16. Do any of our assumptions strike you as worth challenging? There are three things that strike me. First, what does the "ns" mean in Figure 8.16a? It turns out that there is a nonspecific band that shows up when IκB-α is immunoprecipitated (by itself, not together with NF-κB). This cross-reactivity is a major problem with antibodies, as they often bind proteins in addition to the protein of interest. In this case, the authors had to verify that they had the actual band they wanted, and then they had to ensure that their image analysis did not include the ns bands.

How about the fit to the line: Do the data in Figure 8.16b look like they fit a line to you? They don't to me; I would be more likely to choose a nonlinear fit (maybe an exponential fit, for example). The linear fit may be the simplest, but it is not necessarily the most accurate.

Finally, did the researchers collect enough data points? The plot in Figure 8.15b is used to calculate a half-life of ~550 min, but the data do not actually go out for even half that long. Projecting the data out so far is called **extrapolation**, and it can be problematic. For example, even though the plot in Figure 8.15b looks relatively linear for the given time span, based on the plot in Figure 8.16b, you could imagine that the line in Figure 8.15b may look more nonlinear at later time points.

I want to emphasize that I have the highest respect for these experimental scientists; they know what they're doing! Nonetheless, as a future model builder, it is important for you to think critically about even the best "gold standard" parameters reported in the literature, such as these. This consideration is one of the reasons that I decided to become an experimental biologist (a related reason was the cruel humor of my experimentalist colleagues; see sidebar 8.1).

SIDEBAR 8.1   A COMPUTATIONAL BIOLOGIST WALKS INTO A BAR ...

There's something that every aspiring computational biologist should know: Experimental biologists like to make fun of and joke about us. Here's a great joke: A computational biologist is driving through the countryside when he sees a shepherd with a big flock. The computational biologist pulls over and hops out to talk with the shepherd.

"That's an impressive flock you have there."

"Thanks, Mister."

"Hey, what would you say to a bet? If I can tell you how many sheep are in the group, can I have one?"

"Well, that would be pretty impressive—you've got yourself a deal!"

"Okay, you've got 237 sheep."

"Wow, nicely done! Go ahead and pick one out."

So, the computational biologist grabs one and is heading back to his car when the shepherd calls out: "Say Mister, I'll tell you what—if I can guess what you do for a living, can I have it back?"

The computational biologist says, "I doubt you'll be able to, but sure—you're on."

"You must be a computational biologist."

Flabbergasted, the computational biologist asks, "How on earth did you know that?"

Smiling, the shepherd responds, "Well, that's my dog you're holding."

The point is, if you don't know the biology, you won't have a lot of street cred; the best way to pick up the biology is to start doing experiments.

I wanted to have a better sense of how experiments work and where data come from because I believed that my modeling practice would be better as a result, and this turned out to be true. However, no single lab could have determined all of the parameters that went into the Hoffmann–Levchenko model; many labs and person-decades of work were required. You should therefore recognize that to build a model, you will have to rely on the work of many researchers and many research groups.

BOUNDING PARAMETER VALUES

Six additional parameters of the Hoffmann–Levchenko model were drawn from published ranges instead of published values. These parameters are "bounded" by the literature and include the nuclear import rates of IκB, NF-κB, and IκB:NF-κB (three parameters), the nuclear exports of all the complexes (two parameters), and the degradation rate constant for IκB mRNA.

Let's consider the nuclear import rate of IκB in more detail. In this case, IκB-α was fused to a fluorescent protein, just like TetR (Figure 3.1). Normally, IκB-α resides almost exclusively in the cytoplasm, although it shuttles back and forth to the nucleus just like NF-κB (Figure 8.12). By blocking nuclear export with the export inhibitor leptomycin B and tracking the amounts of fluorescence in the nucleus over time, the nuclear import rate was determined (Figure 8.17a). The fluorescence rises rapidly and then hits a steady state (Figure 8.17b). To calculate the import rate, we fit a line to the first few time points; the slope is the change in fluorescence over time.

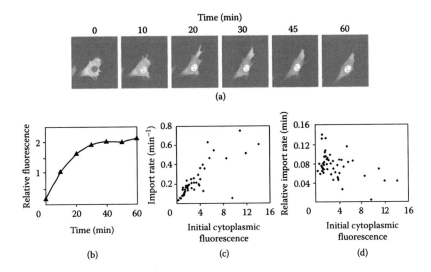

FIGURE 8.17   Cells are treated with leptomycin B, an inhibitor of nuclear export, to determine the nuclear import rate of IκB. (a) Confocal microscopy is used to visualize the translocation of fluorescent IκB-α from the cytoplasm to the nucleus. (b) Quantitation of the IκB-α concentration in the nucleus. (c) Variation in the rate of import of IκB-α into the nucleus as a function of the initial fluorescent IκB-α concentration in the cytoplasm. (d) Relative import rates following treatment with leptomycin B. (From Carlotti, F., Dower, S. K., and Qwarnstrom, E. E. *The Journal of Biological Chemistry* 2000, **275**(52): 41028–41034. © The American Society for Biochemistry and Molecular Biology.)

Notice in Figure 8.17c that the change in fluorescence over time is a function of the initial fluorescence of the cytoplasm. We therefore divide by that initial fluorescence to obtain a normalized nuclear import rate (Figure 8.17d). One look should tell you why this parameter's value is reported as a bounded range: The values range from just under 0.01 to over 0.12! This range gives us something to work with when we are trying to identify the best values for use in our model, and I'll write more about parameter fitting later in this chapter.

MODEL SENSITIVITY TO PARAMETER VALUES

Finally, we have six parameters that were undetermined in the literature in 2002: the constitutive or basal mRNA synthesis rate of all IκB species (three parameters total), the inducible synthesis rate constant for IκB-α, the translation rate of the IκB mRNAs, and the adaptation of the cell to active IKK (meaning the rate at which the potency of active IKK decreases over time; Figure 8.13c).

When you're trying to specify the values of unknown parameters, you might ask what kind of an effect changing that parameter has on the behavior of the model. Some parameters have big effects on the model output (in which case, you need your guesses to be as accurate as possible), while other parameters have small effects (in which case it doesn't matter whether you accurately guess the parameter value).

For example, let's take a look at the output of the Hoffmann–Levchenko model, specifically the prediction it makes for the concentration of NF-κB in the nucleus over time. We will call this the "baseline output," and it is shown as black lines in Figure 8.18.

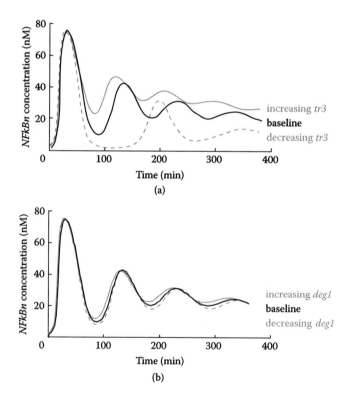

FIGURE 8.18 Comparing simulation outputs enables the modeler to fine-tune their guesses about parameter values. (a) Versus the baseline output (black line), decreasing (dashed gray line) or increasing (gray line) the IκB transcript degradation rate (*tr3*) by 50% dramatically affects NF-κB behavior. (b) Versus the baseline output (black line), decreasing (dashed gray line) or increasing (gray line) the rate of constitutive degradation of free IκB (*deg1*) by 50% has relatively little effect.

As we discussed previously, NF-κB shuttles back and forth between the cytoplasm and the nucleus (Figure 8.9). I should interject here that the baseline output from the Hoffmann–Levchenko model (Figure 8.18) exhibits something like the oscillations that we see in single cells. However, this output does not exhibit pure oscillations; instead, **damping** of the oscillatory behavior is apparent, with each peak less distinct than its predecessor. The damped model was the best fit to the population-level data that Hoffmann and Levchenko had in 2002. Our more recent measurements in individual cells (Figure 8.9) allowed us to update the model.

Let's use the baseline output of this model to determine how much the output changes when we vary a particular parameter. For example, if we increase the IκB transcript degradation rate $tr3$ by 50%, the baseline output changes pretty dramatically (gray line in Figure 8.18a). A higher second peak appears earlier than the second peak in the baseline output, and the period of the oscillation seems to change. In the case of a 50% reduction in $tr3$, the second peak appears much later, and NF-κB makes a much more dramatic exit (dashed gray line in Figure 8.18a).

By comparison, changing the value of the rate of constitutive degradation of free IκB (*deg1*; Figure 8.18b) has little effect on the baseline output. Based on these results, we would say that the baseline output is more sensitive to the value of $tr3$ than to that of *deg1*.

It can be useful to quantify sensitivity with a metric to indicate the degree of difference between two simulations. Here, I just use something simple: If we consider each of the model outputs as a vector of time points, we can calculate the Euclidean distance between two vectors as:

$$\sum_i \left(baseline_i - changed_i\right)^2 \qquad (8.24)$$

We then can represent the total effect of raising or lowering the parameter value by 50% as:

$$\text{Total effect} = \text{Euclidean}(baseline, raised) + \text{Euclidean}(baseline, lowered) \qquad (8.25)$$

In Figure 8.19, the calculation was performed for every parameter in the NF-κB model. Notice that for all but eight or nine of the parameters, a 50% change in the value made essentially no difference! The parameters with

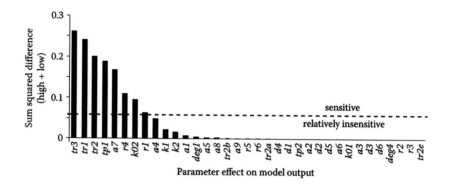

FIGURE 8.19   Total effect of raising or lowering each individual parameter in the Hoffmann–Levchenko model. The effect is calculated using Equation 8.25; parameter names are as in Figures 8.10–8.13. The parameters are listed in the order of highest effect, and a dashed line separates the parameters to which the model is most sensitive (defined as having an effect greater than 20% of the maximum effect, exhibited by *tr3*) from the parameters to which the model is relatively insensitive.

the highest sensitivity are located in the main feedback loop of the system (Figures 8.10–8.13): expression of IκB-α and the interaction of IκB-α with NF-κB and IKK.

Based on this result, you may assume that we do not need to specify the values of certain parameters (the ones for which we calculated a low sensitivity) as concretely as the others. This assumption is not strictly true because a change in the value of one parameter can lead to a change in the sensitivity of the model output to another parameter value. The 50% change that we tried may also not be large enough for the critical parameters to reveal themselves. Nevertheless, you can clearly see that certain parameters have a dominant effect on the model output, which can help to guide you as you start to change parameter values to "fit" your experimental data.

REDUCING COMPLEXITY BY ELIMINATING PARAMETERS

Having performed our sensitivity analysis (Figure 8.19), we can reexamine the six unbounded parameters and ask: Does the model's baseline output change much depending on the values of these six parameters? Unfortunately, the answer is yes. While the baseline output varies little with changes in the basal mRNA expression parameters *tr2a*, *tr2b*, and *tr2e*, the values of the inducible synthesis rate constant for IκB-α (*tr2*), the translation rate of the IκB mRNAs (*trl*), and the adaptation of the cell to active IKK (*k02*) all strongly affect baseline output (Figure 8.19). How do we estimate these six parameters?

Hoffmann and Levchenko used an indirect approach to estimate these parameters by generating NF-κB nuclear localization data (the experimental counterpart to the model's baseline output) and then choosing a set of parameter values that best modeled the experimental data. To understand their investigation, we first need to know how to measure the amount of active NF-κB in the nucleus. Figure 8.20a contains raw data from an **electrophoretic mobility shift assay** (EMSA; sometimes called a "gel shift"). Cell nuclei were isolated and then literally

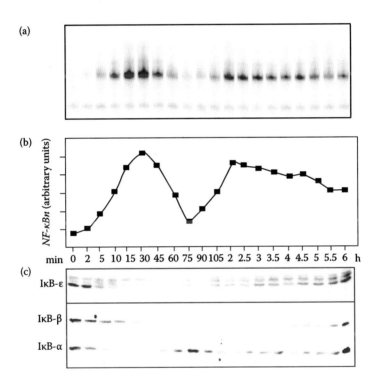

FIGURE 8.20 EMSA allows quantification of the binding of a nuclear extract containing NF-κB to radiolabeled oligonucleotides carrying the NF-κB binding site. (a) Raw data in the form of the radioactive bands on the polyacrylamide gel allow (b) quantitation of NF-κB binding to its binding site on radioactive DNA. (c) Western blotting of the corresponding cytoplasmic fractions reveals the amounts of the IκB isoforms in the cytoplasm. (From Hoffmann, A., Levchenko, A., Scott, M. L., and Baltimore, D. *Science* 2002, **298**: 1241–1245. Reprinted with permission from AAAS.)

cracked open with dry ice to obtain a nuclear extract containing active NF-κB. An oligonucleotide containing a strong NF-κB binding site was radiolabeled and incubated with the nuclear extract, where it was bound by NF-κB. The product was run on a polyacrylamide gel to enable specific visualization of the NF-κB/oligonucleotide complex using a phosphorimager. Higher amounts of active NF-κB lead to darker and larger radioactive bands; the intensity of the band can be quantified (Figure 8.20b). The amounts of IκB were also monitored via western blotting (Figure 8.20c).

Notice that the data in Figure 8.20b do not look exactly like the data in Figure 8.9b, which were taken from individual cells. There are several potential reasons for this discrepancy; most notably, data differ when taken from a single cell versus a population of cells whose behavior is not completely synchronized. In Figure 8.20b, the initial behavior of NF-κB in the pool of cells is sufficiently synchronized to reveal the oscillatory behavior of NF-κB (remember, NF-κB must have been present in the nucleus to bind the radioactive DNA in the EMSA), but this behavior damps out after two peaks of activity.

The goal of Hoffmann et al. was to use these data to fit their model, but with so many loosely specified or unspecified parameter values, I think it became clear to them that data from wild-type extracts were not going to be sufficient. As a result, they pulled off an impressive feat: By a series of molecular modifications and cross-breeding, they produced "double-knockout" mice in which two of the three IκB isoforms had been removed from the mouse's genome. In other words, Hoffmann et al. created mice that only had one IκB isoform. They then performed the same EMSA procedure on cells from these (embryonic) mice (Figure 8.21a,b). It should be immediately apparent to you that IκB-α is responsible for the oscillatory behavior of NF-κB; only the IκB-β and IκB-ε double-knockout cells exhibited oscillations.

The data obtained from the double knockouts simplifies parameter estimation because several of the parameters are not used in each of the double-knockout cells. For example, in the IκB-β and IκB-ε double knockout, you can set the values of *tr2b* and *tr2e* to zero and thus focus on the IκB-α behavior. Hoffmann et al. relied on a combination of intuition and random search: They tried a bunch of parameter values, and they determined which values fit best. By "bunch", I mean many, many, many thousands. But, what do I mean by "intuition"? The original paper states

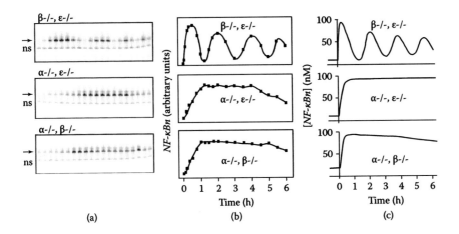

FIGURE 8.21 EMSA of nuclear extracts from three double-knockout strains of mice reveals the driver of NF-κB translocation into the nucleus. Each strain harbors only one of the three IκB isoforms: α (top), β (middle), or ε (bottom). -/- indicates that the deletions were engineered into both strands of the DNA. (a) Raw EMSA data for NF-κB binding to a radiolabeled oligonucleotide. Time points are as in Figure 8.20. Arrows point to the NF-κB band. ns, spots that are not specifically bound to the NF-κB dimer. (b) Quantification of the EMSA data. (c) NF-κB nuclear localization modeling output. (From Hoffmann, A., Levchenko, A., Scott, M. L., and Baltimore, D. *Science* 2002, **298**: 1241–1245. Reprinted with permission from AAAS.)

that "the parameters were determined by a semi-quantitative fitting." That phrase probably means that the random search did not perform very well, so the researchers tweaked parameters by hand to find values that yielded a better fit between the model output and the experimental data. Is there anything fishy about this? No, it just means that progress still remains to be made in parameter estimation for complex biological systems! Also, it turns out that humans still outperform computers in many matters of judgment, reminding us that modeling is an art as much as a science, and that all parameter values should be viewed with optimistic skepticism.

PARAMETER INTERACTIONS

Hoffmann et al. estimated the previously unspecified parameter values for each of the double-knockout strains, resulting in model output that fit the double-knockout data well. For instance, note that the oscillatory pattern of the nuclear NF-κB concentration over time in the IκB-β and IκB-ε double knockouts appears in both model and experimental data; the data from the other double knockouts also resemble the model outputs (Figure 8.21c).

However, when the investigators combined all parameters in their wild-type model, they no longer captured the wild-type behavior due to interaction(s) among the IκB-related feedback parameters. Increasing the feedback parameters for any of the three IκB isoforms will increase the total potency of IκB and thereby reduce the activation of NF-κB; simultaneously increasing the parameters for all three of these isoforms completely overwhelms the circuit (Figure 8.22a).

To counteract this interaction, the basal transcription rates for IκB-β and IκB-ε (*tr2b* and *tr2e*, respectively) were reduced by sevenfold, leading to a damped oscillation that was similar to the experimental behavior of NF-κB in wild-type cells (Figure 8.22b). Hoffmann and Levchenko also judged that a further fivefold decrease in these parameters changed the model too far in the opposite direction, resulting in an overly oscillatory response (Figure 8.22c).

Finally, Hoffmann et al. compared their model output to all of their experimental data (Figure 8.23). The EMSA data (Figure 8.20a,b) did not fit perfectly, but the data captured the main behavior: damped oscillation. Similarly, the western blots revealed the oscillating concentration of IκB-α, as well as the rapid decrease and slower recovery of the IκB-β and IκB-ε concentrations (Figure 8.20c).

In this chapter, we have moved deep into the heart of one of the best computational models in systems biology. This model has guided the field of NF-κB dynamics for over a decade and has led to many important new insights about the system. I hope you come away with a sense of its complexity and how difficult it was to put together!

At the same time, you've explored many of the assumptions that went into its construction. These assumptions should not lead you to criticize this model's validity, but perhaps you have developed an admiration for the experimental, computational, and intellectual effort that went into its construction. Assumptions like these are part and parcel of biological modeling and modeling in general, which is why the statistician George Box said, "All models are wrong, but some are useful." Indeed, the cutting-edge work of Hoffmann and Levchenko et al. has motivated scientists around the world, including myself, to improve the model by testing assumptions, reevaluating parameters, and applying the model's

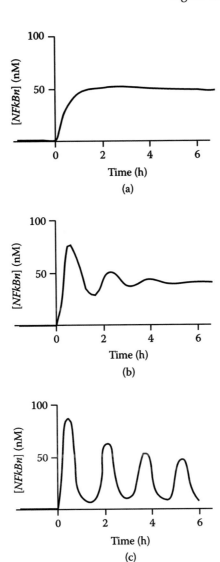

FIGURE 8.22 Moving from the double-knockout parameter values to the wild-type parameter values. (a) NF-κB behavior output when the model uses parameter values from the double-knockout cells. The feedback is too strong for molecular shuttling to occur. (b) Output with a sevenfold reduction in the values for *tr2b* and *tr2e*. This curve is the best match to what we would expect to see from the data. (c) Output with a further fivefold reduction in *tr2b* and *tr2e* exhibits stronger oscillations than the population-level data would suggest. (From Hoffmann, A., Levchenko, A., Scott, M. L., and Baltimore, D. *Science* 2002, **298**: 1241–1245. Reprinted with permission from AAAS.)

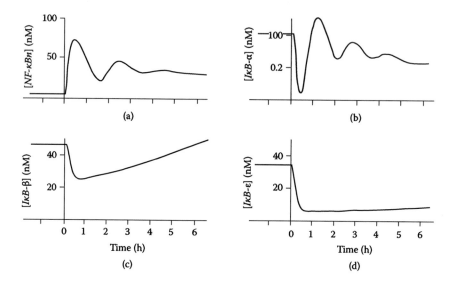

FIGURE 8.23   Final model outputs for various parameters capture the important behavior revealed by the experimental data in Figure 8.20. The model captures the damped oscillatory behavior of NF-κB (a, compare to Figure 8.20a), and the western blots (Figure 8.20c) compare well to predictions for IκB-α (b), IκB-β (c), and IκB-ε (d). (From Hoffmann, A., Levchenko, A., Scott, M. L., and Baltimore, D. *Science* 2002, **298**: 1241–1245. Reprinted with permission from AAAS.)

predictions to different and more complex situations. Similarly, the better you understand how models like these are made, and how the need for certain assumptions and approximations arises, the more useful your own models will be.

CHAPTER SUMMARY

This chapter moved from transcriptional regulation to signal transduction, the process by which cells sense and respond to the outside environment. We began by considering three aspects of signal transduction that we had not treated previously. First, we wrote and analyzed ODEs that described how cellular receptors associate with and dissociate from ligands. Interestingly, the equations that govern these interactions are extremely similar in form to our equations for expression from a constitutively expressed gene in Chapter 3. This similarity means that we can apply many of our tools and much of our intuition to this system.

Adapting our modeling approaches to signal transduction required the inclusion of complexation. We increased the complexity of our models when necessary, then carried out numerical analysis of the behavior of such systems (for example, under steady-state or equilibrium conditions).

Signal transduction nearly always involves translocation events, such as a transcription factor moving to the nucleus or the internalization of a ligand into a cellular endosome. To include translocation, we wrote additional ODEs that described the concentration of the factor in each compartment. Then we added terms to the equations to account for the factor's movement between compartments.

Having touched on these aspects, we next considered the innate immune signaling that leads to NF-κB activation. The original construction and validation of this model was one of the first and most exciting advances in the early days of systems biology. The model itself is large and complex but can be readily understood once the principles of modeling complexation and localization are known.

Finally, practical application of the model forced us to think more carefully about parameters and how to estimate them from experimental measurements. Using the raw data that were available to Hoffmann and Levchenko, we described how they estimated the model's parameters. This discussion included linear least-squares fitting and basic sensitivity analysis, but I hope it also gave you a sense of the challenges of parameter estimation and how important it is to view both data and models with optimistic skepticism.

RECOMMENDED READING

BioNumbers. Home page. http://www.bionumbers.hms.harvard.edu. This website is a database of biological numbers, many of them approximations that are useful for the types of calculations performed in this book.

Carlotti, F., Dower, S. K., and Qwarnstrom, E. E. Dynamic shuttling of nuclear factor kappa B between the nucleus and cytoplasm as a consequence of inhibitor dissociation. *The Journal of Biological Chemistry* 2000, 275(52): 41028–41034.

Hoffmann, A., Levchenko, A., Scott, M. L., and Baltimore, D. The IκB-NF-κB signaling module: temporal control and selective gene activation. *Science* 2002, 298: 1241–1245.

Lauffenburger, D. A. and Linderman, J. *Receptors: Models for Binding, Trafficking, and Signaling.* New York: Oxford University Press, 1996.

Lee, T. K., Denny, E. M., Sanghvi, J. C., Gaston, J. E., Maynard, N. D., Hughey, J. J., and Covert, M. W. A noisy paracrine signal determines the cellular NF-κB response to lipopolysaccharide. *Science Signaling* 2009, 2(93): ra65.

Lemaitre, B., Nicolas, E., Michaut, L., Reichhart, J. M., and Hoffmann, J. A. The dorsoventral regulatory gene cassette spatzle/Toll/cactus controls the potent antifungal response in *Drosophila* adults. *Cell* 1996, **86**(6):973–983.

Pando, M. P. and Verma, I. M. Signal-dependent and -independent degradation of free and NF-kappa B-bound IkappaBalpha. *Journal of Biological Chemistry* 2000, **275**(28): 21278–21286.

PROBLEMS

PROBLEM 8.1
Receptor-Ligand Binding: Understanding the Effects of Ligand Depletion from the Medium

Let's consider a receptor R and a ligand L binding to form a complex C:

$$R + L \xrightleftharpoons[k_{dissoc}]{k_{assoc}} C$$

In our previous analysis of this system, we assumed that the concentration of free ligand was effectively constant. This assumption holds when the number of available ligands is much greater than the number of receptors. However, when the numbers of available ligands and receptors are roughly equal, the assumption breaks down, and we have to explicitly consider the reduction in the number of available ligands as ligands bind to receptors ("ligand depletion").

Let:

$$R_T = \text{total number of receptors per cell}$$

$$R = \text{number of free receptors per cell}$$

$$[L] = \text{ligand concentration}$$

$$C = \text{number of complexes per cell}$$

$$n = \text{number of cells per volume of medium}$$

$$k_{assoc} = \text{association rate constant}$$

$$k_{dissoc} = \text{dissociation rate constant}$$

N_{Av} = Avogadro's number = 6.022 x 10^{23} molecules/mole

$$K_D = k_{dissoc}/k_{assoc}$$

a. Write an ODE describing the change in complex quantity per cell over time for the reaction shown.

b. We assume that R_T, the total number of receptors (free receptors plus complexed receptors) remains constant. Likewise, $[L_T]$, the total concentration of ligand (bound plus free in the medium), remains constant and is simply equal to $[L]_0$, the amount of ligand originally added in the medium. Write the two conservation equations that reflect these assumptions.

c. Use the conservation equations in (b) to remove R and L from the ODE you wrote in (a).

d. Now let's consider the effect of ligand depletion on this ODE, using TNF-α and TNFR as our example. As you saw in Table 8.1, the following parameters have been measured for TNFR in A549 human epithelial cells:

$$R_T = 6.6 \times 10^3 \text{ receptors per cell}$$

$$k_{assoc} = 9.6 \times 10^8 \ (1/\text{min/M})$$

$$k_{dissoc} = 0.14 \ (1/\text{min})$$

$$K_D = 1.5 \times 10^{-10} \text{ M}$$

We will model the binding of TNF-α to TNFR for the case in which 10^5 A549 cells are growing in 1 well of a 96-well plate (volume 0.2 mL).

Use your equation from (c) to derive an expression for C_{eq}, the amount of complex at equilibrium. Then, use MATLAB to plot C_{eq} versus $[L]_0$ spanning from 10^{-12} M to 10^{-8} M (use steps of 10^{-12} M). Describe this plot in your own words.

(e) Now use ode45 in MATLAB to generate two plots showing the change in C over time, using your equation from (c). First, let $[L]_0 = 3.5 \times 10^{-9}$ M and then repeat your analysis with $[L]_0 = 3.5 \times 10^{-11}$ M. How do these two plots differ?

PROBLEM 8.2
Receptor-Ligand Binding: A Stochastic Simulation

In Problem 8.1, you considered the effect of ligand depletion when the amount of ligand in the environment was roughly equal to the amount of receptor. Now let's use stochastic simulations to model even smaller amounts of ligand and receptor. Again, we consider TNF-α binding to TNFR, but this time in a very small environment (one well of a microfluidic chip) that contains only 10 cells total, each of which has 6,600 TNF-α receptors on its surface. We are interested in whether each individual cell receives noticeably different amounts of ligand under varying environmental conditions. For each cell, two reactions can occur (association or dissociation of the receptor and ligand) for a total of 20 reactions. The constants for each cell are:

$$c_{assoc,i} = 0.00001 \text{ (associations/receptor/ligand/min)}$$

$$c_{dissoc,i} = 0.01 \text{ (dissociations/complex/min)}$$

where $i = 1 \ldots 10$ refers to each individual cell.

a. Run five stochastic simulations, starting with initial conditions in which no receptors have been bound, but 100,000 molecules of TNF-α have just been added to the culture medium. Simulate 80,000 time steps, tracking the number of free ligands in the medium, as well as the number of free receptors and complexes on each individual cell. Plot the number of complexes on each of the 10 cells. What do you observe? Are there any major qualitative differences between these plots and the result you obtained in Problem 8.1e?

b. Now run another five simulations, this time starting with only 1,000 molecules of TNF-α in the culture medium. Run the simulation for 1,200 time steps and plot your results. How do these plots differ from your plots for Problem 8.1e? What are the implications of this difference for performing experiments at low ligand concentrations?

PROBLEM 8.3
Linear Least-Squares Fitting

In previous chapters, we described the use of GFP expression as a reporter of transcriptional activity. Here, consider a gene X that is under the control of an inducible promoter controlled by the concentration of a small-molecule inducer (such as aTc), which we will call S. This promoter also drives GFP expression. Our goal is to determine the effect of changing the concentration of S on the expression of GFP. In our experiment, we add S, wait 2 h, and then examine the cells in a microscope. We should see something like this:

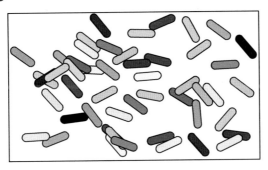

We can then use image analysis software (or MATLAB) to determine the average brightness of each cell. Sample data appear in the following table, for three concentrations of S and 10 cells:

BASELINE AND INDUCED GFP
EXPRESSION (ARBITRARY UNITS)

0 µg/mL S	0.05 µg/mL S	0.1 µg/mL S
1.13	1.51	2.00
1.10	1.69	1.95
1.05	1.26	1.72
0.96	1.48	1.62
0.94	1.55	2.21
0.84	1.31	2.24
0.94	1.57	1.83
1.02	1.23	1.69
0.98	1.55	1.90
1.06	1.71	2.26

a. Does baseline GFP expression (0 µg/mL S) significantly differ from expression at either concentration of S? Use the MATLAB functions ttest and hist to justify your answer (be sure to read the help documents for these functions if you are unfamiliar with these statistical analyses).

b. Determine the mean and standard deviation for each sample and calculate a linear least-squares fit between the mean GFP concentrations and the concentration of S. Do not use a built-in MATLAB function.

c. Generate a scatter plot of all the points and the line found by regression. How could you incorporate this relationship into an ODE describing the production of GFP? (Hint: you can assume that the protein is stable and therefore neglect loss.)

PROBLEM 8.4
Modeling NF-κB Dynamics

We now examine NF-κB activation in the Hoffmann–Levchenko model. As discussed, the structure and parameters in the model were determined from previous experimental measurements and from new experiments performed by the authors. Here, we demonstrate some of the main steps that the authors carried out to generate their model.

The chapter material should be sufficient for you to complete this problem. That said, the citation for this article appears in the Recommended Reading list (Hoffmann et al., 2002), and I recommend that you download and familiarize yourself with it (particularly the supplemental material) if you would like to better understand some of the nuances of the model that I didn't cover in this chapter.

a. Recall that some of the equations in this model are fairly lengthy. The compartment diagrams we discussed in Chapter 3 are useful for keeping track of which proteins bind to which others. The illustration below is a compartment model for the concentration of IκB-α in the cytoplasm. From this diagram, write the ODE describing the cytoplasmic concentration of IκB-α.

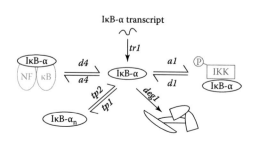

b. Draw a compartment diagram of a model for cytoplasmic NF-κB using the equation from the solution to Practice Problem 8.2. Since this diagram can become crowded, you don't need to draw the IκB-β and IκB-ε terms, but be sure to understand how they fit into your model.

c. For the next several parts of this problem, you are provided two MATLAB files. The first file, NFkB_eq_set.m, holds most of the equations for the NF-κB model; this code also appears at the end of this problem. Two ODEs have been omitted: the one for cytoplasmic IκB-α and the one for cytoplasmic NF-κB. Add these equations to the file in place of the text %TYPE EQUATION HERE. Remember to include all IκB isoforms in your NF-κB equation!

d. The second MATLAB file, NFkB_solver.m, is a function that will run your NF-κB model; it contains all initial conditions and constants. The code, which also appears at the end of this problem, already includes lines to plot the dynamics of nuclear NF-κB. Add code to plot the dynamics of baseline IκB-α, IκB-β, and IκB-ε over time (three separate graphs) where you see the comment %ADD CODE FOR ADDITIONAL PLOTS HERE. Note that "baseline" means all of the IκB-α, IκB-β, and IκB-ε proteins in the cell, including all free and bound species in the cytoplasm and the nucleus. You will therefore be plotting a sum.

Which of the IκBs exhibits oscillatory behavior? Are NF-κB and the IκBs in or out of phase with each other? What does this mean in biological terms? Will the dynamics you observe at early times persist forever? Why or why not?

e. So far, our NF-κB system does not include a ligand or ligand receptor. TNF-α must bind to three TNFR receptors for maximum activation, but to simplify our model, assume that it only binds to one, and that once bound, the ligand-receptor pair activates IKK:

| TNFR | TNF-α | Complex |
| (R) | (L) | (C) |

Write an ODE for the change in C over time. Apply conservation equations but assume a constant total ligand concentration (no ligand depletion). Let R_T be the total number of receptors per cell and $[L_0]$ be the initial/total ligand concentration.

f. Now let's link C to the activation of IKK (technically, this skips a few steps in the pathway, but that's okay—modelers simplify like this all the time). Let's add a species to the model called $\text{IKK}_{inactive}$ (recall that the IKK term in Hoffmann's original model is always active). Assume that the rate constant for IKK activation by the complex is k_{act} and depends (by mass action kinetics) on the concentrations of complex and inactive IKK. The rate of IKK inactivation depends only on the amount of active IKK with a rate constant of k_{inact}.

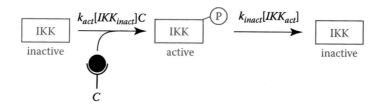

The activation of IKK does not cause the TNF-α:TNFR complex (C) to dissociate, and you can assume that all of the IKK in the original model is and remains active (the active IKK that dissociates from complexes with NF-κB and IκB remains active). In other words, IKK inactivation will only occur based on the new reactions that you add to the model as described previously. Write ODEs for the changes in the concentrations of active and inactive IKK over time.

g. Let's integrate our model with your equations from (e) and (f), together with the following constants and initial conditions (add these to your code):

```
%constants
kdissoc = 0.1;  %1/min
kinact = 0.1;  %1/min
kassoc = 10^-8;  %1/uM*min
kact = 0.5;  %1/uM*min
RT = 10^4;  %receptors/cell
L0 = 100;  %uM
```

```
%initial conditions
IKK0 = 0; %this now represents active IKK
IKKinact0 = 0.1;
complex0 = 0;
```

Remember that in Hoffmann and Levchenko's original model, they considered all of the IKK to be active. Thus, whenever you encounter "IKK" in the original code, it refers to active IKK. Also, note that in the ODE for the active IKK, you will need to add the terms in the ODE you developed in (f) to the existing terms in the code (including complexation, signal degradation, etc.). Run your new code and graph the concentrations of baseline NF-κB, IκB-α, IκB-β, and IκB-ε, as well as active IKK, inactive IKK, and the TNF-α:TNFR complex C over time for 400 min. As the concentration of the complex reaches steady state, what happens to the IKK concentrations?

h. Your current model required the addition of several new parameters, and it is important to know how much the model output depends on these parameter values. To determine the model sensitivity to k_{assoc}, rerun your code for $k_{assoc} = \{10^{-5}, 10^{-6}, 10^{-7}, 10^{-8}, 10^{-9}, 10^{-10}\}$ $1/(\mu M \cdot min)$. Plot the results for baseline NF-κB concentration versus time on the same set of axes. Now, perform a separate sensitivity analysis for k_{act}, beginning with the original k_{assoc} value and rerunning your code for $k_{act} = \{0.2, 0.4, 0.6, 0.8, 1\}$ $1/(\mu M \cdot min)$. Again, plot the results for baseline NF-κB concentration versus time on the same set of axes. What do you observe?

NFkb_eq_set.m

```
function dy = NFkB_eq_set(t,y,c)
% this contains all of the equations for
% the Hoffmann-Levchenko model.

% put the constants back in their names
a1 = c(1);   a2 = c(2);   a3 = c(3);   a4 = c(4);
a5 = c(5);   a6 = c(6);   a7 = c(7);   a8 = c(8);
a9 = c(9);

d1 = c(10);  d2 = c(11);  d3 = c(12);  d4 = c(13);
d5 = c(14);  d6 = c(15);
```

```
deg1 = c(16); deg4 = c(17);

k01 = c(18); k02 = c(19);
k1 = c(20); k2 = c(21);

r1 = c(22); r2 = c(23); r3 = c(24); r4 = c(25);
r5 = c(26); r6 = c(27);

tp1 = c(28); tp2 = c(29);
tr1 = c(30); tr2 = c(31);
tr2a = c(32); tr2b = c(33); tr2e = c(34);
tr3 = c(35);

% The y array is as follows:
% y(1)  = IkBa            y(2)  = IkBa_NFkB
% y(3)  = IkBan           y(4)  = IkBan_NFkBn
% y(5)  = IkBat           y(6)  = IkBb
% y(7)  = IkBb_NFkB       y(8)  = IkBbn
% y(9)  = IkBbn_NFkBn     y(10) = IkBbt
% y(11) = IkBe            y(12) = IkBe_NFkB
% y(13) = IkBen           y(14) = IkBen_NFkBn
% y(15) = IkBet           y(16) = IKK
% y(17) = IKK_IkBa        y(18) = IKK_IkBa_NFkB
% y(19) = IKK_IkBb        y(20) = IKK_IkBb_NFkB
% y(21) = IKK_IkBe        y(22) = IKK_IkBe_NFkB
% y(23) = NFkB            y(24) = NFkBn
% y(25) = fr

dy(1) = %TYPE EQUATION HERE
dy(2) = a4*y(1)*y(23) -d4*y(2) ...
        -a7*y(16)*y(2) +d1*y(18) +k2*y(4) ...
        -deg4*y(2);
dy(3) = tp1*y(1) -tp2*y(3) -a4*y(3)*y(24) ...
        +d4*y(4);
dy(4) = a4*y(3)*y(24) -d4*y(4) -k2*y(4);
dy(5) = tr2a +tr2*y(24)^2 -tr3*y(5);
dy(6) = -a2*y(16)*y(6) +d2*y(19) ...
        -a5*y(6)*y(23) +d5*y(7) ...
        +tr1*y(10) -deg1*y(6) -tp1*y(6) ...
        +tp2*y(8);
dy(7) = a5*y(6)*y(23) -d5*y(7) ...
        -a8*y(16)*y(7) +d2*y(20) ...
        +0.5*k2*y(9)*y(25) -deg4*y(7);
```

```
dy(8)   = tp1*y(6) -tp2*y(8) -a5*y(8)*y(24) ...
          +d5*y(9);
dy(9)   = a5*y(8)*y(24) -d5*y(9) ...
          -0.5*k2*y(9)*y(25);
dy(10)  = tr2b -tr3*y(10);
dy(11)  = -a3*y(16)*y(11) +d3*y(21) ...
          -a6*y(11)*y(23) +d6*y(12) +tr1*y(15) ...
          -deg1*y(11) -tp1*y(11) +tp2*y(13);
dy(12)  = a6*y(11)*y(23) -d6*y(12) ...
          -a9*y(16)*y(12) +d3*y(22) ...
          +0.5*k2*y(14) -deg4*y(12);
dy(13)  = tp1*y(11) -tp2*y(13) ...
          -a6*y(13)*y(24) +d6*y(14);
dy(14)  = a6*y(13)*y(24) -d6*y(14) -0.5*k2*y(14);
dy(15)  = tr2e -tr3*y(15);
dy(16)  = -k02*y(16) -a1*y(16)*y(1) ...
          +(d1+r1)*y(17) -a2*y(16)*y(6) ...
          +(d2+r2)*y(19) -a3*y(16)*y(11) ...
          +(d3+r3)*y(21) -a7*y(16)*y(2) ...
          +(d1+r4)*y(18) -a8*y(16)*y(7) ...
          +(d2+r5)*y(20) -a9*y(16)*y(12) ...
          +(d3+r6)*y(22);
dy(17)  = a1*y(16)*y(1) -(d1+r1)*y(17) ...
          -a4*y(17)*y(23) +d4*y(18);
dy(18)  = a7*y(16)*y(2) +a4*y(17)*y(23) ...
          -(d1+d4+r4)*y(18);
dy(19)  = a2*y(16)*y(6) -(d2+r2)*y(19) ...
          -a5*y(19)*y(23) +d5*y(20);
dy(20)  = a8*y(16)*y(7) +a5*y(19)*y(23) ...
          -(d2+d5+r5)*y(20);
dy(21)  = a3*y(16)*y(11) -(d3+r3)*y(21) ...
          -a6*y(21)*y(23) +d6*y(22);
dy(22)  = a9*y(16)*y(12) +a6*y(21)*y(23) ...
          -(d3+d6+r6)*y(22);
dy(23)  = %TYPE EQUATION HERE
dy(24)  = k1*y(23) -a4*y(3)*y(24) +d4*y(4) ...
          -a5*y(8)*y(24) +d5*y(9) ...
          -a6*y(13)*y(24) +d6*y(14) -k01*y(24);
dy(25)  = -.5/(1 + t)^2;

dy      = [dy(1);dy(2);dy(3);dy(4);dy(5);
          dy(6); dy(7); dy(8); dy(9); dy(10);
          dy(11);dy(12);dy(13);dy(14);dy(15);
```

```
            dy(16);dy(17);dy(18);dy(19);dy(20);
            dy(21);dy(22);dy(23);dy(24);dy(25)];
```

NFkB_solver.m

```
function NFkB_solver()

% Here are all the default constants
% a1: Association between IKK and IkBa (No NFkB)
% a2: Association between IKK and IkBb (No NFkB)
% a3: Association between IKK and IkBe (No NFkB)
% a4: Association between IkBa and NFkB
% a5: Association between IkBb and NFkB
% a6: Association between IkBe and NFkB
% a7: Association between IKK and IkBa-NFkB
% a8: Association between IKK and IkBb-NFkB
% a9: Association between IKK and IkBe-NFkB

a1 = 1.35;              % 1/(uM*min)
a2 = 0.36;              % 1/(uM*min)
a3 = 0.54;              % 1/(uM*min)
a4 = 30;                % 1/(uM*min)
a5 = 30;                % 1/(uM*min)
a6 = 30;                % 1/(uM*min)
a7 = 11.1;              % 1/(uM*min)
a8 = 2.88;              % 1/(uM*min)
a9 = 4.2;               % 1/(uM*min)

% d1: Dissociation between IKK and IkBa (No NFkB)
% d2: Dissociation between IKK and IkBb (No NFkB)
% d3: Dissociation between IKK and IkBe (No NFkB)
% d4: Dissociation between IkBa and NFkB
% d5: Dissociation between IkBb and NFkB
% d6: Dissociation between IkBe and NFkB

d1 = 0.075;             % 1/min
d2 = 0.105;             % 1/min
d3 = 0.105;             % 1/min
d4 = 0.03;              % 1/min
d5 = 0.03;              % 1/min
d6 = 0.03;              % 1/min

% deg1: Rate of constitutive degradation of free IkB
% deg4: Rate of constitutive degradation of IkB-NFkB
```

```
% k01:   Transport rate of nuclear NFkB to cytoplasm
% k02:   Attenuation of the IKK signal
% k1:    Transport rate of cytoplasmic NFkB to nucleus
% k2:    Transport rate of IkB-NFkB to the nucleus


deg1 = 0.00675;       % 1/min
deg4 = 0.00135;       % 1/min
k01 = 0.0048;         % 1/min
k02 = 0.0072;         % 1/min
k1 = 5.4;             % 1/min
k2 = 0.82944;         % 1/min


% r1: Reaction rate between IKK and IkBa (No NFkB)
% r2: Reaction rate between IKK and IkBb (No NFkB)
% r3: Reaction rate between IKK and IkBe (No NFkB)
% r4: Reaction rate between IKK and IkBa-NFkB
% r5: Reaction rate between IKK and IkBb-NFkB
% r6: Reaction rate between IKK and IkBe-NFkB


r1 = 0.2442;          % 1/min
r2 = 0.09;            % 1/min
r3 = 0.132;           % 1/min
r4 = 1.221;           % 1/min
r5 = 0.45;            % 1/min
r6 = 0.66;            % 1/min


% tp1: Rate of IkB nuclear import
% tp2: Rate of IkB nuclear export
% tr1: IkB isoform mRNA transcript translation rate
% tr2: Enhancement rate of NFkB on IkBa transcription
% tr2a: Constitutive rate of IkBa transcription
% tr2b: Constitutive rate of IkBb transcription
% tr2e: Constitutive rate of IkBe transcription
% tr3:  Degradation rate of IkB transcripts


tp1 = 0.018;          % 1/min
tp2 = 0.012;          % 1/min
tr1 = 0.2448;         % 1/min
tr2 = 0.99;           % 1/min
tr2a = 9.21375E-05;   % uM/min
tr2b = 0.000010701;   % uM/min
tr2e = 0.000007644;   % uM/min
tr3 = 0.0168;         % 1/min
```

```
defaultConstants = ...
[a1; a2; a3; a4; a5; a6; a7; a8; a9;
 d1; d2; d3; d4; d5; d6; deg1; deg4;
 k01; k02; k1; k2; r1; r2; r3; r4; r5; r6;
 tp1; tp2; tr1; tr2; tr2a; tr2b; tr2e; tr3];

% Initial conditions
IkBa0 = 0.185339;               % uM
IkBa_NFkB0 = 0.0825639;         % uM
IkBan0 = 0.18225;               % uM
IkBan_NFkBn0 = 0.00140031;      % uM
IkBat0 = 0.0054872;             % uM
IkBb0 = 0.0212451;              % uM
IkBb_NFkB0 = 0.00934616;        % uM
IkBbn0 = 0.021088;              % uM
IkBbn_NFkBn0 = 0.000313095;     % uM
IkBbt0 = 0.000636964;           % uM
IkBe0 = 0.0151759;              % uM
IkBe_NFkB0 = 0.00667621;        % uM
IkBen0 = 0.0150637;             % uM
IkBen_NFkBn0 = 0.000223652;     % uM
IkBet0 = 0.000455;              % uM
IKK0 = 0.1;                     % uM
IKK_IkBa0 = 0;                  % uM
IKK_IkBa_NFkB0 = 0;             % uM
IKK_IkBb0 = 0;                  % uM
IKK_IkBb_NFkB0 = 0;             % uM
IKK_IkBe0 = 0;                  % uM
IKK_IkBe_NFkB0 = 0;             % uM
NFkB0 = 0.000256516;            % uM
NFkBn0 = 0.000220118;           % uM
fr0 = 0.5;                      % unitless

initial_conditions = ...
[IkBa0; IkBa_NFkB0; IkBan0; IkBan_NFkBn0;
IkBat0; IkBb0; IkBb_NFkB0; IkBbn0; IkBbn_NFkBn0;
IkBbt0; IkBe0; IkBe_NFkB0; IkBen0; IkBen_NFkBn0;
IkBet0; IKK0; IKK_IkBa0; IKK_IkBa_NFkB0; IKK_IkBb0;
IKK_IkBb_NFkB0; IKK_IkBe0; IKK_IkBe_NFkB0; NFkB0;
NFkBn0; fr0];

% baseline wildtype conditions
options = odeset('RelTol',1e-4);
```

```
[t,y]=ode45(@NFkB_eq_set,[0:0.5:380],...
        initial_conditions,options,defaultConstants);

nfkbn_baseline = y(:,24)+y(:,9);

figure(1)
plot(t,nfkbn_baseline);
title('NFkBn baseline');
xlabel('Time (min)');
ylabel('Concentration (uM)');

%ADD CODE FOR ADDITIONAL PLOTS HERE
```

CHAPTER 9

Metabolism

- Derive the Michaelis–Menten equation for enzyme kinetics

- Adapt the equation to account for inhibitory effects

- Build large-scale metabolic models based on principles of flux balance and linear optimization

- Apply flux balance analysis to predict the metabolic behavior of cells

This is an exciting chapter for me; it not only features some of the most classical and well-established modeling in molecular biology, but also showcases some of the newest and most exciting modeling approaches that played a critical role in building a whole-cell model, as you will see in Chapter 10. The focus is metabolism, specifically carbon and energy metabolism, meaning the processes by which a cell builds all of the molecules necessary to grow and divide.

CELLULAR METABOLISM

For example, *E. coli*'s metabolic network enables each individual cell to create a nearly perfect copy of itself. This copy includes more than two million proteins, a DNA molecule over four and a half megabase pairs long, and specialized components like the cell membrane and the flagellum (Figure 9.1). All of these components are created from an external environment containing only glucose and a handful of salts (sodium phosphate, potassium

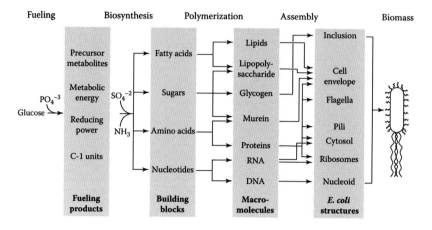

FIGURE 9.1 Schematic of the higher-level structure of the metabolic network of *E. coli*. The general reaction categories appear at the top of the figure, and the general output categories appear at the bottom. Note the hierarchical dependence of this metabolic network; the "upstream" components are required to generate the "downstream" components. (Modified with permission from Neidhardt, F. C., Ingraham, J. L., and Schaechter, M. *Physiology of the Bacterial Cell: A Molecular Approach.* Sunderland, MA: Sinauer Associates, 1990. Copyright © 1992 International Union of Biochemistry and Molecular Biology, Inc.)

phosphate, magnesium sulfate, sodium chloride, ammonium chloride, calcium chloride), all within ~1 h (sometimes as quickly as 20 min!).

First, several chemical reactions break down glucose and phosphate to produce free carbon, energy sources such as **ATP (adenosine triphosphate)**, and **reducing power** (a store of electrons that can be donated to chemical reactions). If you have taken a biochemistry course, you know these reactions as **glycolysis** and the **tricarboxylic acid (TCA) cycle** (also known as the Krebs cycle or the citric acid cycle). The products of these reactions, together with sulfate and ammonium, react to produce the four major building blocks of the cell: fatty acids, sugars, amino acids, and nucleotides. These building blocks come together in various ways to produce the macromolecules of the cell, which in turn comprise the major specialized structures and, ultimately, *E. coli* itself (Figure 9.1).

METABOLIC REACTIONS

Let's look at one of the reactions in more detail to obtain a sense of its components (Figure 9.2). First, we see the molecular reactants and products: phosphoenolpyruvate (PEP), ADP (adenosine diphosphate), and a

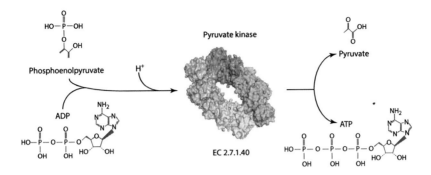

FIGURE 9.2   The components of the pyruvate kinase reaction. The two small molecules phosphoenolpyruvate and ADP react, with the addition of a hydrogen ion, to form the products pyruvate and ATP. This reaction depends on the enzyme pyruvate kinase. Note the EC number below the enzyme, which provides a unique categorization of the reaction regardless of the host organism or even which enzyme catalyzes it.

hydrogen ion react to form pyruvate and ATP. This reaction requires the catalyzing power of the enzyme pyruvate kinase. Two pyruvate kinase genes are encoded in the *E. coli* genome, and the functional protein is built of four of either of these gene products. When an enzyme and the reactants are present in the cell, a chemical reaction can proceed (Figure 9.2).

One major difference between metabolism and the other processes we studied in this book is that the enzymes, reactants, and reactions are incredibly well conserved between organisms. We share most of our metabolic network even with microbes—an observation that led the Nobel Laureate Jacques Monod to assert, "Anything found to be true of *E. coli* must also be true of elephants." However, to translate what we learn from parasites to pachyderms, we need to form consistent connections between what we learn in each organism.

Fortunately for us, the biochemistry community has already adopted a single standard notation to describe reactions, whether they occur in *Triponema* or the tarantula: the **Enzyme Commission** (EC) number, which uniquely specifies a reaction in a hierarchical fashion. For example, the pyruvate kinase reaction EC number is 2.7.1.40. The "2" means that pyruvate kinase is a transferase; it transfers a phosphate group (the "7") to an alcohol group (the "1") with the specific reactants and products shown in Figure 9.2 (the "40"). Because these reactions generally occur in many organisms and cells, even though the genes and corresponding proteins that catalyze the reactions may not be conserved, the EC number is a convenient way to compare metabolic capacity between organisms. You'll soon see that it becomes useful for model building as well.

These reactions were studied individually by biochemists over many years, but for our purposes, we are especially interested in putting them in context, first in pathways and then in networks. Pyruvate kinase, for example, is at the end of the glycolysis pathway, a fueling reaction (Figure 9.3).

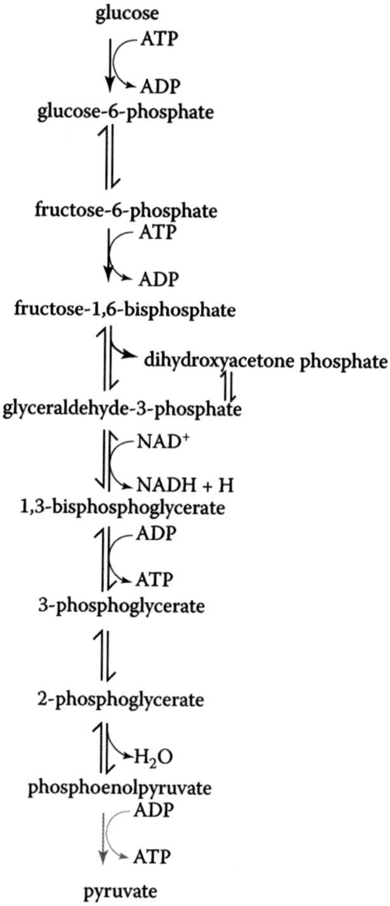

glucose
ATP
ADP
glucose-6-phosphate

fructose-6-phosphate
ATP
ADP
fructose-1,6-bisphosphate

dihydroxyacetone phosphate
glyceraldehyde-3-phosphate
NAD⁺
NADH + H
1,3-bisphosphoglycerate
ADP
ATP
3-phosphoglycerate

2-phosphoglycerate

H₂O
phosphoenolpyruvate
ADP
ATP
pyruvate

FIGURE 9.3 The pyruvate kinase reaction is highlighted in the context of the glycolytic pathway in *E. coli*. The glycolytic pathway begins with glucose as a substrate, which is processed through a number of chemical reactions, each catalyzed by one or more enzymes in the cell, until its eventual conversion to pyruvate via pyruvate kinase. Note from the arrows that some reactions are bidirectional; others only proceed in the forward direction.

This pathway starts with the six-carbon glucose molecule, which is charged with two phosphate groups and converted to fructose 1,6-bisphosphate, after which it is split in half and used to generate four charged ATP molecules and reducing power in the form of NADH (nicotinamide adenine dinucleotide, reduced).

Glycolysis is only one of the many pathways in the full *E. coli* metabolic network (Figure 9.4). Some of the pathways are mostly linear, such as the amino acid biosynthesis pathways; others are cyclic, such as the energy-generating TCA cycle. A substantial fraction of the reactions involve multiple reactants, and a handful of **metabolites** are involved in the lion's share of the reactions (such as ATP). Note that we did not observe long pathways or cycles as motifs in the transcriptional regulatory network (Chapter 7).

COMPARTMENT MODELS OF METABOLITE CONCENTRATION

How can we model such a complicated network? We start with the standard mass balance compartment models used in Chapter 3. Let's say that you are interested in the concentration of a particular metabolite M in the cell; we will call $d[M]/dt$ the change in the metabolite pool. Just as with all of the compartment models we built previously, there are some ways for the metabolite to be produced, and other ways for it to be lost (Figure 9.5). The metabolite concentration increases as reactions produce it or decreases as the metabolite is used as a reactant in other reactions. The metabolite concentration can also increase or decrease as it is transported into or out of the cell.

If we put all of these together, we obtain

$$\frac{d[M]}{dt} = v_{prod} - v_{loss} + v_{trans,in} - v_{trans,out} \tag{9.1}$$

THE MICHAELIS–MENTEN EQUATION FOR ENZYME KINETICS

How do we represent the terms in Figure 9.5? Based on what we have already covered, you may first try a mass action-based approach, like $v_{loss} = k_{Mloss} \cdot [M]$, or possibly $v_{loss} = k_{Mloss} \cdot [M] \cdot [E]$, where E is the enzyme that uses metabolite M. However, this strategy does not explicitly account for the mechanism of enzyme–substrate binding. Consider the following

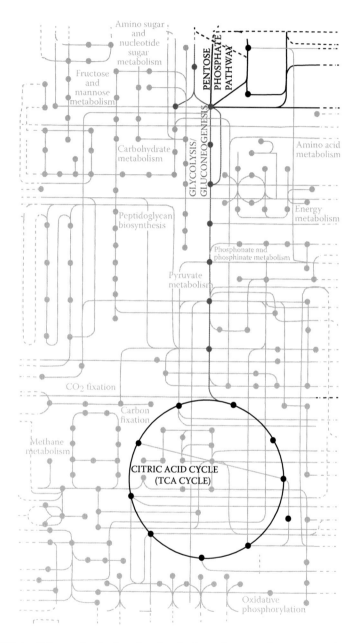

FIGURE 9.4 Schematic placing the glycolytic pathway in the context of the entire metabolic network of *E. coli*. The small black circles represent small molecules, and the lines represent chemical reactions. The glycolytic pathway is highlighted in red. The dashed lines indicate that this complex network extends well beyond what is shown here. (Figure generated using SEED: Overbeek, R. et al. *Nucleic Acids Research* 2005, **33**(17): 5691–5702).

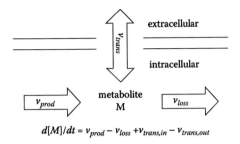

$$d[M]/dt = v_{prod} - v_{loss} + v_{trans,in} - v_{trans,out}$$

FIGURE 9.5 The compartment model for the change in concentration of a metabolite. A metabolite M can be produced or lost by chemical transformations or transported into or out of the cell through channels in the cell membrane. The rates of production, loss, and transport are given by v_{prod}, v_{loss}, and v_{trans}, respectively.

chemical reaction between an enzyme and a single substrate to form a product:

$$E + S \underset{k_r}{\overset{k_f}{\rightleftharpoons}} C \xrightarrow{k_{cat}} E + P \tag{9.2}$$

Here, enzyme E and substrate S bind to form a complex C, which can dissociate back to E and S or react to form a new product P and release the free enzyme. Notice that three kinetic parameters characterize this system: k_f and k_r for association and dissociation, respectively, and k_{cat} for the reaction. Note also that the second reaction is considered irreversible.

You should now be able to write out the four equations that describe the conservation of E, S, C, and P:

$$\frac{d[S]}{dt} = k_r[C] - k_f[S][E] \tag{9.3}$$

$$\frac{d[E]}{dt} = (k_r + k_{cat})[C] - k_f[S][E] \tag{9.4}$$

$$\frac{d[C]}{dt} = k_f[S][E] - (k_r + k_{cat})[C] \tag{9.5}$$

$$\frac{d[P]}{dt} = k_r[C] \tag{9.6}$$

We can also define the initial conditions that $[E](t = 0) = [E]_0$, $[S](t = 0) = [S]_0$, and $[C](t = 0) = [P](t = 0) = 0$.

Look at Equations 9.3–9.6. Which one seems to correspond to the v_{prod} (or v_{loss}) that we are trying to derive? The answer is Equation 9.6, $d[P]/dt$. We want to determine the rate at which product is formed by the reaction. Since we know that this rate is equal to $k_{cat}[C]$, we only need to solve Equations 9.3–9.5 for $[S]$, $[E]$, and $[C]$ to find our answer.

Next, take a closer look at $d[E]/dt$ (Equation 9.4) and $d[C]/dt$ (Equation 9.5). Do you notice anything? One equation is the opposite of the other:

$$\frac{d[E]}{dt} = -\frac{d[C]}{dt} \tag{9.7}$$

or

$$\frac{d[E]}{dt} + \frac{d[C]}{dt} = 0 \tag{9.8}$$

Since the change in the concentration of E will lead to an equal and opposite change in the concentration of C, the sum of $[E]$ and $[C]$ at any time must equal the sum of $[E]$ and $[C]$ at any other time—including time zero. This relationship yields another interesting conservation statement:

$$[E](t) + [C](t) = [E](t=0) + [C](t=0) \tag{9.9}$$

or, because $[C](t=0) = 0$:

$$[E](t) = [E]_0 - [C](t) \tag{9.10}$$

With Equation 9.10, we no longer need to solve both Equation 9.4 and Equation 9.5, just one of them! As a result, we solve only these two equations:

$$\frac{d[S]}{dt} = k_r[C] - k_f[S][E] \tag{9.11}$$

$$\frac{d[C]}{dt} = k_f[S][E] - (k_r + k_{cat})[C] \tag{9.12}$$

To help us solve these equations efficiently, we need to do a little bit of variable definition and substitution. Let's define a term r as the ratio of complex to initial enzyme concentration:

$$r = \frac{[C]}{[E]_0} \qquad (9.13)$$

or

$$[C] = [E]_0 r \qquad (9.14)$$

Substituting into Equation 9.11, we obtain:

$$\frac{d[S]}{dt} = k_r [E]_0 r - k_f [S][E] \qquad (9.15)$$

Remembering that $[E](t) = [E]_0 - [C](t) = [E]_0 - [E]_0 \cdot r$, we substitute to find:

$$\frac{d[S]}{dt} = k_r [E]_0 r - k_f [S]([E]_0 - [E]_0 r) = [E]_0 \left(k_r r - k_f (1-r)[S] \right) \quad (9.16)$$

Now, let's take a look at Equation 9.12, where we can make a number of similar substitutions to obtain a new equation for r:

$$\frac{dr}{dt} = \left(\frac{1}{[E]_0} \right) \frac{d[C]}{dt} = k_f [S] \frac{([E]_0 - [C])}{[E]_0} - (k_r + k_{cat}) \frac{[C]}{[E]_0} \qquad (9.17)$$

$$= k_f (1-r)[S] - (k_r + k_{cat}) r$$

If you compare Equation 9.16 with Equation 9.17, you'll see that their forms are similar. Both equations have a $k_f(1 - r)[S]$ term, and both have a $k_r \cdot r$ term, for example. The most notable difference between the two equations is that the entire right side of Equation 9.16 is multiplied by $[E]_0$. This difference is something we might be able to use to further simplify our model. Can we make any assumptions about $[E]_0$? One useful assumption is that the amount of enzyme is much smaller than the amount of substrate, so $[E]_0 \ll [S]$; since $[C]$ is also always a fraction of $[E]_0$, we can expand this assumption to say that $[C] \ll [S]$ as well.

Following this reasoning, we can also see that $d[S]/dt$, with its E_0 term, must be much smaller than dr/dt, so small that we can treat it as negligible in the context of the system:

$$\frac{d[S]}{dt} \approx 0 \tag{9.18}$$

Put another way, $[S](t) \approx [S]_0$. This simplification is exciting because now we have represented the entire system in one single equation:

$$\frac{dr}{dt} = k_f(1-r)[S]_0 - (k_r + k_{cat})r \tag{9.19}$$

which can be readily solved analytically (Chapter 3) and graphically (Chapter 4), as you will see in the following practice problem.

PRACTICE PROBLEM 9.1

Determine the behavior of Equation 9.19 by plotting dr/dt versus r. Draw a vector field on the plot and identify and characterize any fixed points in terms of their stability.

SOLUTION

We want to solve for r, so first we rearrange Equation 9.19, grouping terms with an r and those without:

$$\frac{dr}{dt} = k_f[S]_0 - (k_r + k_{cat} + k_f[S]_0)r \tag{9.20}$$

Recognizing that $k_f[S]_0$ and $(k_r + k_{cat} + k_f[S]_0)$ are constants will help you to see that this problem is similar to things you encountered in Chapter 4. For example, the plot of r versus dr/dt is a straight line (Figure 9.6), similar to what you saw in Figure 4.3.

The y intercept is $k_f[S]_0$. The fixed point is shown in Figure 9.6, and because the entire vector field points toward it, the fixed point is stable. The value of the fixed point r_{eq} is calculated at $dr/dt = 0$:

$$r_{eq} = \frac{k_f[S]_0}{k_r + k_{cat} + k_f[S]_0} \tag{9.21}$$

With this solution, it should be easy for you to plot the dynamic response of r to multiple initial conditions.

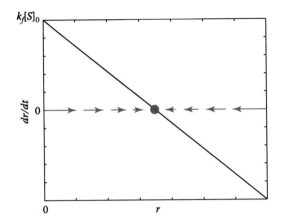

FIGURE 9.6   Graphical analysis of Equation 9.20. Note the vector field and the fixed point.

To graph the dynamic response of C, remember that $r = [C]/[E]_0$; therefore, $r_{eq} = [C]_{eq}/[E]_0$, leading to the following equation for $[C]_{eq}$:

$$[C]_{eq} = \frac{k_f [S]_0 [E]_0}{k_r + k_{cat} + k_f [S]_0} \qquad (9.22)$$

By defining a new quantity, the constant K_m, we can simplify Equation 9.21. We define K_m as the ratio between the loss and production constants for E in the overall system:

$$K_m = \frac{k_r + k_{cat}}{k_f} \qquad (9.23)$$

You can think of K_m as indicating the enzyme's affinity for the substrate. A high K_m means that the dissociation of enzyme and substrate—whether passively (k_r) or through catalysis of a chemical reaction to form product (k_{cat})—is favored over association (k_f). With K_m defined, we can rearrange our expression for $[C]_{eq}$:

$$[C]_{eq} = \frac{k_f [S]_0 [E]_0}{k_r + k_{cat} + k_f [S]_0} \cdot \frac{1/k_f}{1/k_f} = \frac{[S]_0 [E]_0}{\dfrac{k_r + k_{cat}}{k_f} + [S]_0} = \frac{[S]_0 [E]_0}{K_m + [S]_0} \qquad (9.24)$$

This equilibrium concentration of the complex gives us a basis for the production of product. When the system is at equilibrium, we can substitute $[C]_{eq}$ into Equation 9.6 to obtain:

$$\frac{d[P]}{dt} = k_{cat}[C]_{eq} = \frac{k_{cat}[E]_0[S]_0}{K_m + [S]_0} = k_{cat}[E]_0\left(\frac{[S]_0}{K_m + [S]_0}\right) \qquad (9.25)$$

You can mentally partition this equation into two parts: one within parentheses and one outside. The part within parentheses is a ratio that will always fall between zero and one. If the initial substrate concentration is much smaller than K_m, then the ratio will be close to zero and the rate of product production will be very small. On the other hand, if the substrate concentration is much larger than K_m, then the ratio is close to one, meaning that the rate is essentially unrestricted by the substrate concentration. At $[S]_0 = K_m$, the ratio is equal to ½, so the production rate is one-half of what it would be if substrate were plentiful.

You might have deduced from the preceding paragraph that the maximum rate of product production is equal to the part of Equation 9.24 outside the parentheses, $k_{cat}E_0$; this quantity is often called the maximum rate, or V_{max}, so that Equation 9.25 becomes:

$$\frac{d[P]}{dt} = \frac{V_{max}[S]_0}{K_m + [S]_0} = v \qquad (9.26)$$

where v is the rate of product formation (you should be saying, "Aha! A v!" Yes, I'm finally getting back to Figure 9.5). This equation might look familiar because it is often called the Michaelis–Menten model of enzyme kinetics.

A plot of Equation 9.26 appears in Figure 9.7. As discussed previously, at $[S]_0 = K_m$, the rate of product formation is one-half of the maximal rate; in other words, $v([S]_0 = K_m) = ½ \cdot V_{max}$. Furthermore, at low $[S]_0$, the rate of formation is well represented by simple mass action kinetics—the substrate concentration multiplied by a constant (V_{max}/K_m)—while at high $[S]_0$, the rate approaches the maximum value. This scenario suggests that, under some conditions, it might be acceptable to use a constant (high $[S]_0$) or a simple proportionality term (low $[S]_0$) to model the rate of product formation.

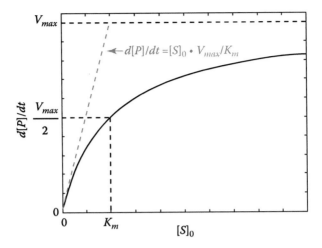

FIGURE 9.7   Graphical representation of Equation 9.26.

DETERMINING KINETIC PARAMETERS FOR THE MICHAELIS–MENTEN SYSTEM

Using the Michaelis–Menten framework, we can experimentally determine the parameters K_m and V_{max} for a given enzyme by introducing a known amount of enzyme and substrate and measuring product formation. I want to give you a real-world example of how this determination works, so let's take a behind-the-scenes look at some data gathered in my lab. We once analyzed the following chemical reaction:

deoxyuridine + phosphate ion ↔ deoxyribose-1-phosphate + uracil

which is catalyzed by the enzyme pyrimidine-nucleoside phosphorylase (EC 2.4.2.2). To determine the K_m and V_{max} for this enzyme, and assuming that the concentration of phosphate ion is typically not limiting in the cell, we constructed samples with differing concentrations of deoxyuridine in addition to a constant amount of enzyme and the appropriate buffers. After a set period, we stopped the reaction (with sodium hydroxide in this case) and measured the amount of uracil via **spectrophotometry** (uracil can be detected by illumination at A_{290}). Dividing the amount of uracil by the length of time, we obtained v, the rate of uracil production. A plot of deoxyuridine concentration versus v generated in my lab is shown in Figure 9.8. Notice that the shape of these data roughly resembles the Michaelis–Menten curve (they are real data, however, so yes, there are outliers!).

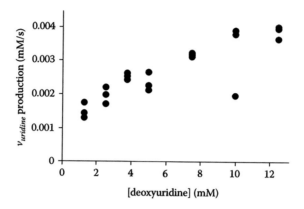

FIGURE 9.8 Deoxyuridine concentration versus the rate of uridine production. Experimental details appear in the main text. (From Sanghvi, J. C., Regot, S., Carrasco, S., Karr, J. R., Gutschow, M. V., Bolival, B., Jr., and Covert, M. W. *Nature Methods* 2013, **10**(12): 1192–1195.)

But how do we derive the Michaelis–Menten parameters from these data? You have already learned how to perform a linear least-squares fit, but in our case we do not have a line to work with.

Or do we? The classical method for determining V_{max} and K_m was developed by Hanes and Woolf, who formulated a linear representation of the Michaelis–Menten equation. First, they inverted Equation 9.26:

$$\frac{1}{v} = \frac{K_m + [S]_0}{V_{max}[S]_0} \tag{9.27}$$

Next, they multiplied both sides by $[S_0]$ and rearranged the fraction:

$$\frac{[S]_0}{v} = \frac{K_m + [S]_0}{V_{max}} = \frac{1}{V_{max}}[S]_0 + \frac{K_m}{V_{max}} \tag{9.28}$$

Notice that Equation 9.28 describes a linear relationship between $[S]_0$ and $[S]_0/v$, plotted for our data as shown in Figure 9.9. Fitting the data to a line, just as we did in Chapter 8, yields the y intercept and the slope, which give us K_m/V_{max} and $1/V_{max}$, respectively (Figure 9.9a). In our case, the V_{max} turns out to be 0.00437 mM/s, and the K_m is 3.13 mM (you can calculate these values directly from the slope and intercept in Figure 9.9a). Using these numbers, we draw the Michaelis–Menten curve that best describes the original data in Figure 9.8 (Figure 9.9b).

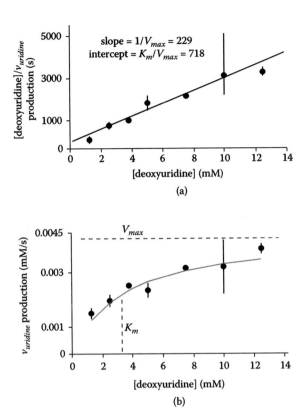

FIGURE 9.9 Determining the parameters for Equation 9.28. (a) The Hanes–Woolf plot; the points are the averages of the data in Figure 9.8, and a line is fit to them. (b) The parameters from (a) are used to draw the corresponding Michaelis–Menten curve (red). (From Sanghvi, J. C., Regot, S., Carrasco, S., Karr, J. R., Gutschow, M. V., Bolival, B., Jr., and Covert, M. W. *Nature Methods* 2013, **10**(12): 1192–1195.)

These days, there are newer, nonlinear computational regression methods that can obtain more accurate determinations of K_m and V_{max}, but in many cases, the parameter values determined via these methods are indistinguishable from those obtained via the Hanes–Woolf method. Moreover, I want you to see how some mathematical gymnastics can enable you to use your linear parameter estimation methods even in the case of a nonlinear problem.

INCORPORATING ENZYME INHIBITORY EFFECTS

The Michaelis–Menten model is based on a number of assumptions that we have described, but an additional assumption that we have not mentioned bears further investigation. Specifically, this model assumes that the interaction between enzyme and substrate is unregulated, although it is well known that metabolic networks are extensively controlled. For example, in Chapter 7 we discussed regulation at the transcriptional level, using the arginine biosynthesis regulon (Figure 7.16) as an example.

There are other ways to regulate enzyme activity at much faster timescales, four of which are shown in Figure 9.10. During substrate inhibition, excess substrate itself prevents effective enzyme–substrate interaction. In some cases, an inhibitor molecule can form a complex with the enzyme to prevent optimal substrate binding. Direct binding between the inhibitor and the enzyme's active site is called competitive inhibition. The inhibitor may also act by binding a different part of the enzyme, thereby triggering a structural change in the enzyme that changes the active site, a process known as noncompetitive inhibition. A final type of regulation is called uncompetitive inhibition, when the inhibitor binds the enzyme–substrate complex and prevents release of the substrate from the enzyme.

We'll investigate competitive inhibition here (Figure 9.11); I'll leave it to you to apply these principles to other types of regulation.

The ODEs that we will use to solve this system are:

$$\frac{d[S]}{dt} = k_r[C] - k_f[E][S] \tag{9.29}$$

$$\frac{d[E]}{dt} = k_r[C] + k_{cat}[C] + k_{-i}[D] - k_f[E][S] - k_i[E][I] \tag{9.30}$$

$$\frac{d[C]}{dt} = k_f[E][S] - k_r[C] - k_{cat}[C] \tag{9.31}$$

$$\frac{d[I]}{dt} = k_{-i}[D] - k_i[E][I] \tag{9.32}$$

$$\frac{d[D]}{dt} = k_i[E][I] - k_{-i}[D] \tag{9.33}$$

$$\frac{d[P]}{dt} = k_{cat}[C] \tag{9.34}$$

We will adopt the same assumptions and use an approach that is similar to the way we tackled the simpler system without regulation. First,

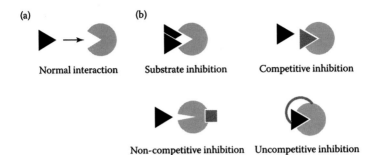

FIGURE 9.10 **Strategies for regulating enzyme activity.** (a) Normal interactions between enzyme (circle) and substrate (triangle). The substrate enters the enzyme's active site (notch in the circle), and catalysis proceeds. (b) Four types of regulation. Inhibitor molecules are depicted in red (inhibitors may have structures similar to that of the substrate or to a conformation of the substrate during catalysis).

$$E + S \underset{k_r}{\overset{k_f}{\rightleftharpoons}} C \overset{k_{cat}}{\longrightarrow} E + P$$

$$+$$

$$I$$

$$k_i \Big\Vert k_{-i}$$

$$D$$

FIGURE 9.11 The set of reactions we will use to explore competitive inhibition. E = enzyme; S = substrate; C = complex; P = product; D = a "disabled" complex of enzyme and inhibitor. The forward (k_f), reverse (k_r), catalytic (k_{cat}), inhibitory (k_i), and reverse inhibitory (k_{-i}) rate constants are shown.

we remember that given the relatively large concentration of S, $d[S]/dt$ is roughly equal to zero in comparison to the other processes; therefore, $[S] \approx [S]_0$. We make an analogous assumption about the concentration of inhibitor I, so that $[I] \approx [I]_0$. Next, we again assume that the total amount of enzyme does not change over time, so that $[E]_0 = [E](t) + [C](t) + [D](t)$. Finally, again we will focus on equilibrium solutions, where $d[E]/dt = d[C]/dt = d[D]/dt = 0$.

Application of these assumptions leads to a simplified set of equations. Let's first consider the simplified version of Equation 9.31 for $d[C]/dt$:

$$\left. \frac{d[C]}{dt} \right|_{[C]_{eq},[E]_{eq},[D]_{eq}} = 0 = k_f [E]_{eq} [S]_0 - (k_r + k_{cat})[C]_{eq} \qquad (9.35)$$

Solving for $[E]_{eq}$, we obtain:

$$[E]_{eq} = \frac{(k_r + k_{cat})[C]_{eq}}{k_f [S]_0} = K_m \frac{[C]_{eq}}{[S]_0} \tag{9.36}$$

The simplified ODE for $d[D]/dt$ (Equation 9.33) becomes:

$$\left. \frac{d[D]}{dt} \right|_{[C]_{eq},[E]_{eq},[D]_{eq}} = 0 = k_i [E]_{eq} [I]_0 - k_{-i} [D]_{eq} \tag{9.37}$$

By defining a new constant $K_I = k_{-i}/k_i$ (analogous to K_m), we can solve for $[D]_{eq}$:

$$[D]_{eq} = \frac{k_i [E]_{eq} [I]_0}{k_{-i}} = \frac{[E]_{eq} [I]_0}{K_I} \tag{9.38}$$

Substituting for $[E]_{eq}$ leads to:

$$[D]_{eq} = \frac{[I]_0}{K_I} \frac{K_m [C]_{eq}}{[S]_0} = \frac{K_m}{K_I} \frac{[I]_0 [C]_{eq}}{[S]_0} \tag{9.39}$$

We can then substitute into our total enzyme conservation equation at equilibrium:

$$[E]_0 = [C]_{eq} + [E]_{eq} + [D]_{eq} \tag{9.40}$$

$$= [C]_{eq} + K_m \frac{[C]_{eq}}{[S]_0} + \frac{K_m}{K_I} \frac{[I]_0 [C]_{eq}}{[S]_0} \tag{9.41}$$

$$= [C]_{eq} \left(1 + \frac{K_m}{[S]_0} + \frac{K_m}{K_I} \frac{[I]_0}{[S]_0} \right) \tag{9.42}$$

Solving for $[C]_{eq}$, we find:

$$[C]_{eq} = [E]_0 \bigg/ \left(1 + \frac{K_m}{[S]_0} + \frac{K_m}{K_I} \frac{[I]_0}{[S]_0} \right) \tag{9.43}$$

$$= [E]_0 [S]_0 \bigg/ \left([S]_0 + K_m + \frac{K_m}{K_I} [I]_0 \right) \tag{9.44}$$

Then, $d[P]/dt$ (Equation 9.34) becomes:

$$\frac{d[P]}{dt} = k_{cat}[C]_{eq} \tag{9.45}$$

$$= k_{cat}[E]_0[S]_0 \bigg/ \left([S]_0 + K_m + \frac{K_m}{K_I}[I]_0\right) \tag{9.46}$$

$$= V_{max}[S]_0 \bigg/ \left([S]_0 + K_m + \frac{K_m}{K_I}[I]_0\right) \tag{9.47}$$

$$= V_{max}[S]_0 \bigg/ \left([S]_0 + K_m\left(1 + \frac{[I]_0}{K_I}\right)\right) \tag{9.48}$$

So, what happened here? What is the difference between Equation 9.48 and the Michaelis–Menten equation (Equation 9.26)? The answer is that K_m in the denominator is now multiplied by a factor that is greater than or equal to one, depending on the ratio between the concentration of the inhibitor and the inhibitor dissociation constant K_I. As shown in Figure 9.12, if the ratio between $[I]_0$ and K_I is 1, then K_m is multiplied by 2, enough to significantly reduce the rate of product formation if $[S]_0$ is comparable to or less than K_m. Increasing $[I]_0$ relative to K_I further decreases the product formation rate; a low $[I]_0$ relative to K_I makes the system essentially analogous to the Michaelis–Menten system without regulation (Figure 9.7).

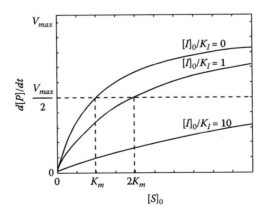

FIGURE 9.12 Competitive inhibition at various values of $[I]_0/K_I$. The V_{max} for all of these values remains the same (eventually, all of these curves will approach the top of the y axis), but the effective K_m increases significantly with an increase in inhibitor, as shown by the dashed lines.

FLUX BALANCE ANALYSIS

The Michaelis–Menten approach to modeling enzyme kinetics is now used widely in a variety of biological systems. It has also been built upon, by adding the framework for two substrates. Larger models generally incorporate multiple Michaelis–Menten types of systems. However, some scientists have emphasized that in many cases the approximations required to derive the Michaelis–Menten equation may not always be valid. In such cases, we simply rely on Equations 9.3–9.6, which represent the changes in concentration for S, E, C, and P, respectively.

Either way, just as with our transcription and signaling models, a major obstacle and modeling challenge is to determine the necessary parameter values. Determining the kinetic parameters for even a single enzyme can be a significant amount of work, as you saw in the section "Determining Kinetic Parameters for the Michaelis-Menten System," and determining parameters for an entire metabolic network is just plain infeasible with current technology. This obstacle means that it should be extremely difficult to build useful larger-scale ODE-based models of metabolism.

However, since the mid-1990s, a new modeling method has greatly expanded the scope of metabolic models, allowing the creation of models that account for every known metabolic gene and reaction. This method is called **flux balance analysis** (FBA), and it was developed in the laboratory of Bernhard Palsson, now at the University of California, San Diego.

The main idea behind FBA is to use linear optimization to get around the need to have defined kinetic parameters. As an example, let's consider the simple metabolic network in Figure 9.13: A nutrient N is transported into a cell and converted directly into biomass (X) being formed and transported out of the cell (think of biomass as corresponding to new cells). The entire system consists of one metabolic reaction and two transport reactions.

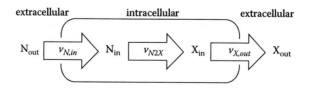

FIGURE 9.13 Our simple metabolic network. The nutrient N is transported into the cell and converted into biomass X, which is "transported out" as the cell grows and divides. This network is characterized by three metabolic processes, two transport processes and one chemical reaction, which are each associated with flux ($v_{N,in}$, $v_{X,out}$, and v_{N2X}, respectively). The N2X notation indicates the conversion of N into X.

To model this system using ODEs, we begin with:

$$\frac{d[M]}{dt} = \sum v_{production} - \sum v_{loss} \tag{9.49}$$

where $[M]$ is the metabolite concentration and v represents the fluxes that arise when one metabolite is converted to another (as in Figure 9.5 and Equation 9.1, neglecting transport terms in this case). You can see from Equation 9.49 that the units of v are concentration over time; typically, concentrations are normalized by the weight of dry cellular biomass in the culture (called the dry cell weight, or DCW) to enable comparisons across time. Thus, in practice, the units of v are millimoles per gram DCW per hour (mmol/g DCW/h).

Note also that the presence of the cellular and extracellular compartments in Figure 9.13 requires us to consider four metabolites because both N and X can be present inside or outside the cell. For our system, the equations are:

$$\frac{d[N_{out}]}{dt} = -v_{N,in} \tag{9.50}$$

$$\frac{d[N_{in}]}{dt} = v_{N,in} - v_{N2X} \tag{9.51}$$

$$\frac{d[X_{in}]}{dt} = v_{N2X} - v_{X,out} \tag{9.52}$$

$$\frac{d[X_{out}]}{dt} = v_{X,out} \tag{9.53}$$

Next, we model and parameterize each flux term. For example, we might assume that the fluxes follow Michaelis–Menten kinetics, and write:

$$\frac{d[N_{in}]}{dt} = \frac{V_{max(N,in)} \cdot [N_{out}]}{[N_{out}] + K_{m(N,in)}} - \frac{V_{max(N2X)} \cdot [N_{in}]}{[N_{in}] + K_{m(N2X)}} \tag{9.54}$$

As you see, we would need to experimentally determine or otherwise specify these four parameters for this equation (six total for the set). In the case of *E. coli*, for which more than 1,000 metabolic reactions have been identified (Figure 9.4), we would need to determine thousands of parameters—not to mention the additional parameters required for inhibition or other regulation!

FBA takes a different tack. Here, rates of change in metabolite concentration determined by fluxes (Equation 9.1) remain represented as fluxes and put into matrix notation:

$$
\begin{bmatrix} \frac{d[N_{out}]}{dt} \\ \frac{d[N_{in}]}{dt} \\ \frac{d[X_{in}]}{dt} \\ \frac{d[X_{out}]}{dt} \end{bmatrix} = \begin{bmatrix} -1 & & \\ +1 & -1 & \\ & +1 & -1 \\ & & +1 \end{bmatrix} \begin{bmatrix} v_{N,in} \\ v_{N2X} \\ v_{X,out} \end{bmatrix}
\tag{9.55}
$$

where the zero values are represented as empty spaces for clarity. Recall from your linear algebra class that the product of an m-by-n matrix and an n-by-1 vector is an m-by-1 vector.

Equation 9.55 can be represented more compactly as:

$$
\frac{d[M]}{dt} = Sv
\tag{9.56}
$$

where $d[M]/dt$ is a vector, S is a two-dimensional matrix, and v is a vector, identical to Equation 9.55. Notice that Equations 9.55–9.56 are also identical to Equations 9.50–9.53.

Let's take a closer look at matrix S in Equation 9.55, considering in turn the rows and the columns:

$$
\begin{array}{c}
\\
N_{out} \\
N_{in} \\
X_{in} \\
X_{out}
\end{array}
\begin{array}{ccc}
v_{N,in} & v_{N2X} & v_{X,out} \\
\end{array}
\begin{bmatrix}
-1 & & \\
+1 & -1 & \\
& +1 & -1 \\
& & +1
\end{bmatrix}
$$

The rows each correspond to a particular metabolite, and you have already seen how the entries correspond to the production and loss terms for that metabolite's ODE.

Each column corresponds to a reaction, and the entries in that column are the stoichiometric coefficients for that reaction. For example, look at the left column, which is linked to the reaction that transports N into the cell. Note that the top entry is in the N_{out} row, and its value is -1; the next entry down is in the N_{in} row, with a value of $+1$. A negative sign corresponds to metabolites that are used (substrates); the positive signs indicate a reaction product. Accordingly, from the left column, we can write:

$$[1]N_{out} \xrightarrow{v_{N,in}} [1]N_{in} \qquad (9.57)$$

This is the reaction itself! So, rather than working out all of the conservation relationships to build the matrix S in terms of its rows, you can write matrix S more simply as a group of reaction stoichiometries. For this reason, we often call matrix S the stoichiometric matrix.

Note that matrix S is invariant with respect to a particular organism because the stoichiometry of reactions does not change; one glucose molecule has exactly six carbon atoms, so at most one fructose molecule (six carbons) could be produced through glycolysis. The invariance of matrix S is markedly different from the other kinetic parameters we have discussed, the determination of which can vary widely, depending on the experimental conditions and analysis, as you saw with NF-κB in Chapter 8. As long as the genome (and consequently, the metabolic network of the cell) remains the same, matrix S will be constant. This scenario has some major advantages for large-scale modeling, which is why the rearrangement of the set of ODEs to obtain matrix S is one of the best features of FBA.

STEADY-STATE ASSUMPTION AND EXCHANGE FLUXES

Now we have one completely specified matrix and two unknown vectors. To simplify the problem, Palsson and colleagues decided to focus on the steady-state solutions. This strategy seemed justified, as most research in the field of metabolic engineering was concerned with observations over the hours or days that it takes cells to divide, as opposed to the timescales required for a metabolic reaction to occur (typically on the order of seconds). Concentrating on the longer timescales is akin to stating that $d[M]/dt \approx 0$, and therefore:

$$Sv = 0 \qquad (9.58)$$

Obtaining this small equation was a huge result because decades of research had already gone into solving this kind of problem (a known matrix multiplied by a vector of unknowns equals a zero vector) using linear optimization. Airlines, oil companies, and shipping companies decide how to route their fleet based on variations of Equation 9.58, and major financial investment companies also use it.

However, there are certain problems with the steady-state assumption; for example, what do we do about chemical species outside the cell? Consider *E. coli* growing and taking up glucose, for example, or even the simplified example in Figure 9.13. In this case, the rate of change of the glucose concentration in the outside environment must be nonzero, so $d[M]/dt$ for that row would have a nonzero value.

Palsson and colleagues addressed this issue by introducing the concept of an **exchange flux**. Figure 9.14 adds two exchange fluxes to our simple network from Figure 9.13. Exchange fluxes, denoted by convention with the letter *b*, appear to be fluxes to nowhere. They are missing substrates or products because they allow metabolites to enter or exit the system.

Using the exchange flux convention, if $d[M]/dt$ for a certain metabolite outside the cell had a nonzero value, then we would not be able to use Equation 9.58. Instead, let's go back to Equation 9.49, which as you remember is a statement of the mass balance for each metabolite. If we say that $d[M]/dt$ is equal to a value *b*, then:

$$b = \sum v_{production} - \sum v_{loss} \tag{9.59}$$

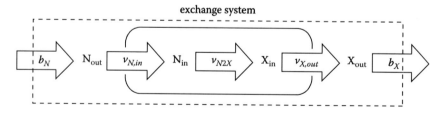

FIGURE 9.14   Adding exchange fluxes to the simple network in Figure 9.13. N_{out} and X_{out} are present outside the cell, and their concentrations change over time (picture a cell feeding on a nutrient and creating biomass). The exchange fluxes enable N_{out} and X_{out} to enter or leave the system, respectively. By convention, exchange fluxes are labeled with a *b* instead of the *v* used for most fluxes.

Rearranging:

$$0 = \sum v_{production} - \sum v_{loss} - b \tag{9.60}$$

We can apply Equation 9.60 to Figure 9.14 by writing the full set of conservation equations:

$$\frac{d[N_{out}]}{dt} : 0 = -v_{N,in} + b_N \tag{9.61}$$

$$\frac{d[N_{in}]}{dt} : 0 = v_{N,in} - v_{N2X} \tag{9.62}$$

$$\frac{d[X_{in}]}{dt} : 0 = v_{N2X} - v_{X,out} \tag{9.63}$$

$$\frac{d[X_{out}]}{dt} : 0 = v_{X,out} - b_X \tag{9.64}$$

Writing these equations in matrix notation yields a new S_b matrix and v_b vector:

$$\begin{bmatrix} \frac{d[N_{out}]}{dt} \\ \frac{d[N_{in}]}{dt} \\ \frac{d[X_{in}]}{dt} \\ \frac{d[X_{out}]}{dt} \end{bmatrix} = \begin{bmatrix} -1 & & & +1 & \\ +1 & -1 & & & \\ & +1 & -1 & & \\ & & +1 & & -1 \end{bmatrix} \begin{bmatrix} v_{N,in} \\ v_{N2X} \\ v_{X,out} \\ b_N \\ b_X \end{bmatrix}$$

or

$$0 = S_b v_b \tag{9.65}$$

In other words, Equation 9.65 is an equivalent way of writing the conservation relationship with the metabolite concentrations such that the left side of the equation is zero.

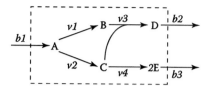

FIGURE 9.15   The network for Practice Problem 9.2.

PRACTICE PROBLEM 9.2

Give the stoichiometric matrix for the system shown in Figure 9.15. Label the matrix rows and columns.

SOLUTION

The matrix is:

$$
\begin{array}{c c}
& \begin{array}{c c c c c c c}
v1 & v2 & v3 & v4 & b1 & b2 & b3
\end{array} \\
\begin{array}{c}
A \\ B \\ C \\ D \\ E
\end{array} &
\left[\begin{array}{c c c c c c c}
-1 & -1 & & & +1 & & \\
+1 & & -1 & & & & \\
& +1 & -1 & -1 & & & \\
& & +1 & & & -1 & \\
& & & +2 & & & -1
\end{array}\right]
\end{array}
$$

Notice that the entries do not have to simply be +1, –1, or zero. They don't even have to be integers, as you'll see later.

SOLUTION SPACES

In linear algebra, the solution to an equation with the form of Equation 9.64 is called the **null space** of S_b (FBA terminology calls this space the **solution space**). The operable word here is "space", illustrated in Figure 9.16a. Usually, the system is **underdetermined**, meaning that there are many possible solutions to the equations. Assuming that all of the fluxes in our sample system can only take on positive values, the positive orthant (gray) in Figure 9.16 holds all possible combinations of flux values that can be attained by the system, both dynamic and at steady state. The null space of S_b is shown in black (but only schematically, for illustrative purposes; in reality, the solution space is just the straight line $v_{N,in} = v_{N2X} = v_{X,out}$). This space holds all of the solutions for which $d[M]/dt = 0$. A particular ODE solution, based

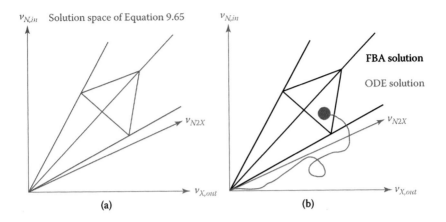

FIGURE 9.16 Schematic example of the solutions to Equation 9.65 produced by FBA (black) or by solving a set of ODEs (red). FBA yields a solution space that contains all of the possible steady-state solutions. The ODE solution contains a single trajectory over time that may eventually reach a steady state. The flux terms are as described in Figure 9.14. The exchange flux axes are not shown to keep this figure two-dimensional.

on determination of the necessary parameter values as well as an initial condition, will follow a trajectory. In the case illustrated in Figure 9.16b, the trajectory begins at the origin and then travels outside the FBA solution space until the system reaches a steady state, which necessarily falls inside the solution space.

THE OBJECTIVE FUNCTION

As mentioned in the section entitled "Steady-state Assumption and Exchange Fluxes", this mathematical approach is used by major companies, and financial analysts and logistics consultants also calculate solution spaces to describe their problems. However, these people are not interested in the entire space. Instead—this is going to shock you—they are interested in making money. They want to identify the solutions in the space that maximize profits or minimize costs. To identify these **optimal solutions**, they define an **objective**. With a bounded solution space and an objective, they can use linear optimization to find their solutions.

We could also apply linear optimization to identify points in the metabolic solution space by defining an objective. But do cells actually have objectives? It sounds more than a little anthropomorphic, and it indeed was a hotly contested assumption when it was first published, but FBA relies on the assumption that the cellular metabolic network can be treated as if it has an objective. Consider an *E. coli* culture growing in a test tube on your bench. This culture comes from a single cell, but it is not uniform. Many mutations have occurred during culture, and some of these mutations could enable one cell to grow a little faster than its neighbors. Let's assume that some cells arise that grow ~10% faster than the rest of the cells in the culture. If those faster-growing cells initially only represent 1% of the total population, they will take over the entire population in less than 1 day (Figure 9.17).

In other words, the fastest-growing cells are highly favored in the lab environment, so under those conditions we might be able to treat them as if they had an "objective" to maximize the amount of biomass they produce. This is the most common objective invoked in FBA.

Can we make this assumption for all kinds of cells? No. An objective of maximizing growth would be totally inappropriate when modeling the metabolic network of a neuron, for example. However, this assumption might be relevant to the growth of a tumor cell, which is characterized by unrestrained growth.

How do we mathematically represent such an objective? Scientists have measured the chemical composition of certain cells, including *E. coli*, from which we can determine the biomass maximization objective.

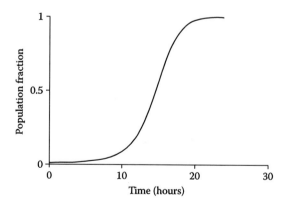

FIGURE 9.17 The population fraction of a group of cells that divides faster (growth rate 0.4 divisions/h) than the rest of the population (growth rate 0.36 divisions/h). This group goes from making up 1% of the population to nearly 100% in less than a day. This graph is based on a theoretical estimation, solving the growth equation $X = X_0 e^{\mu t}$ for both populations.

For example, we know that, under certain conditions, *E. coli* is 70% water and that the remaining dry *E. coli* biomass consists of 55% protein, 20.5% RNA, 3.1% DNA, 9.1% lipid, 3.4% lipopolysaccharide, 2.5% murine, 2.5% glycogen, and 3.9% polyamines, metabolites, cofactors, and ions (by weight), and thus we also know that the metabolic network needs to produce these biomass components in these amounts to make cells. Several of these components can be specified further; for example, 1 g of *E. coli* DCW contains 0.205 g RNA, which has 165 μmol AMP (adenosine monophosphate), 203 μmol GMP (guanosine monophosphate), 126 μmol CMP (cytidine monophosphate), and 136 μmol UMP (uridine monophosphate).

Using these numbers, we can devise a pseudochemical reaction that describes biomass formation. Analogous to the way in which we represent the conversion of PEP, ADP, and hydrogen ion to pyruvate and ATP with the equation PEP + ADP + H$^+$ → pyruvate + ATP, we can also write:

$$\sum Protein\ subunits + \sum RNA\ subunits + \sum DNA\ subunits + \ldots$$
$$\xrightarrow{v_{biomass}} 1\,\text{g dry biomass} \tag{9.66}$$

We can even unpack the summations to read, for example:

$$\sum Protein\ subunits + \left(\begin{array}{c} 165\,\mu\text{mol AMP} + 203\,\mu\text{mol GMP} + 126\,\mu\text{mol CMP} \\ + 136\,\mu\text{mol UMP} \end{array} \right)$$
$$+ \sum DNA\ subunits + \ldots \xrightarrow{v_{biomass}} 1\,\text{g dry biomass} \tag{9.67}$$

The full equation will include dozens of subunits, with coefficients determined from the scientific literature (see Recommended Reading).

Notice that Equation 9.67 is essentially a conversion of molecular concentrations to DCW. This conversion results in a change in the units of the biomass flux: Whereas typical fluxes have units of mmol/g DCW/h, the biomass flux has units of 1/h. In fact, you know another name for the value of this flux: the growth rate!

DEFINING THE OPTIMIZATION PROBLEM

With an objective function defined, we can express the linear optimization problem in terms of that function and a set of constraints that define the solution space:

Maximize (or minimize): _____

Subject to these constraints: _____

_____

_____

Applied to our case, so far we have:

$$\text{Maximize: } v_{biomass}$$

Subject to these constraints: $S_b v_b = 0$ (Equation 9.65, conservation of mass)

This is a good start, but the solution space could be reduced even more, giving us a tighter lock on our answer, if we could come up with other constraints. FBA typically incorporates two other types of constraints, the first of which concerns the reversibility of chemical reactions. Some of the reactions in a metabolic network can go in either direction (paired arrows in Figure 9.11); others can only proceed in one direction (single arrow in Figure 9.11), usually because of the energetics or the enzymes required for the reaction. In our current definition of the problem, the flux values for all of the reactions (v_i) are unconstrained, so:

$$-\infty \le v_i \le +\infty \tag{9.68}$$

However, when a given reaction is known to be irreversible, we can write the constraint as:

$$0 \le v_i \le +\infty \tag{9.69}$$

Adding these reactions generally collapses the solution space through several dimensions, but there is another type of constraint that can reduce it even more. Consider the fact that no real enzyme can actually produce a flux value of infinity. Instead, each enzyme has a maximum capacity, and if we knew that capacity, we could add in that constraint:

$$-V_{max} \leq v_k \leq V_{max} \tag{9.70}$$

Measuring V_{max} is difficult; as you have seen, in most cases we would need to determine k_{cat} for that enzyme as well as the amount of enzyme itself. In general, we only look at certain key transporters to consider how much nutrient the cells take up from the environment and how much product they secrete. We use those numbers to establish our constraints, so now we have:

Maximize: $v_{biomass}$

Subject to these constraints: $S_b v_b = 0$ (conservation of mass)

$$0 \leq v_j \leq +\infty \text{ (irreversibility)}$$

$$-V_{max} \leq v_k \leq +V_{max} \text{ (maximum capacity)}$$

SOLVING FBA PROBLEMS USING MATLAB

All of these constraints can be encoded in MATLAB using the function `linprog`. The function looks like this:

```
>> [vb_vals, obj_val] = linprog(obj_fxn, [], [], Sb,
zero_vals, lower_bounds, upper_bounds);
```

where `vb_vals` is the set of all of the fluxes; `obj_val` is the value of the flux for the objective function only, `obj_fxn` is a vector that contains the objective function (it's written as a reaction column vector, so normally we implement this by adding an exchange flux for the biomass to exit the system and then maximizing that flux; note that `linprog` always minimizes the objective function, so to maximize it, multiply `obj_fxn` by −1), `Sb` is the stoichiometric matrix, `zero_vals` is a vector of zeros (the right side of $S_b v_b = 0$), and `lower_bounds` and `upper_bounds` are vectors that define the low and high bound, respectively, for each flux.

Tip: One issue with `linprog` is that it does not work well for real-world problems (don't ask me why!), so if you're planning on adding FBA to your research tool kit, you'll want to consider better solvers, such as GLPK, which can be integrated with MATLAB. Since GLPK can be tedious to install, `linprog` will be sufficient for our learning and practice here.

Let's put this all together with a real-world example. Let's say that you wanted to build a metabolic model for *E. coli* (the Palsson lab has actually built, tested, updated, and distributed such a model for almost two decades now; see Orth et al., 2011). How would you start?

The first step would be to determine the metabolic genes in *E. coli*. This "parts list" for the network can be obtained from the annotated genome sequence. For example, a table of protein functions can be obtained from the Marine Biological Laboratory (http://genprotec.mbl.edu/files/geneproductfunctions.txt). I've listed the first five entries in Table 9.1. The first two columns (Eck and bnum) contain different types of gene ID. The next column is the gene name. Notice that the first four genes are the *thr* operon, with the leader *thrL* followed by the enzyme-encoding genes *thrA*, *thrB*, and *thrC*. The functions of these genes are presented in the next column, "Gene Type." The final columns provide the names of the gene product and the EC numbers if applicable. Notice that *thrA* encodes a fused enzyme that can catalyze two reactions, and hence it has two EC numbers.

With the gene names and EC numbers, we can attach reactions to ThrA, ThrB, and ThrC and try to organize them into a pathway—or we can take advantage of the website EcoCyc (http://www.ecocyc.org), run by Peter Karp and colleagues. If you go to the EcoCyc website and search for "threonine," then click on the "threonine biosynthesis" pathway page (it should be at the top of the search results), you will see something that looks like Figure 9.18.

There is a lot of information in this representation. First, you can see how L-aspartate is converted to L-threonine via five chemical reactions. Some reactions are reversible, and others are not, as indicated by the arrows. The necessary cofactors and by-products for each reaction are shown, with a list of any genes that encode products that catalyze

TABLE 9.1   Gene Production Function Table for *E. coli* K-12 as of August 28, 2007

Eck	bnum	Gene	Gene Type	Gene Product	EC
ECK0001	b0001	*thrL*	l	thr operon leader	
ECK0002	b0002	*thrA*	e	Fused aspartokinase, homoserine dehydrogenase I	2.7.2.4, 1.1.13
ECK0003	b0003	*thrB*	e	Homoserine kinase	2.7.1.39
ECK0004	b0004	*thrC*	e	Threonine kinase	4.2.3.1
ECK0005	b0005	*yaaX*	o	Predicted protein	

Source: The information in this table is an excerpt from genprotec.mbl.edu/files/geneproductfunctions.txt. l = leader peptide; e = enzyme; o = other. Just worry about the e rows for now.

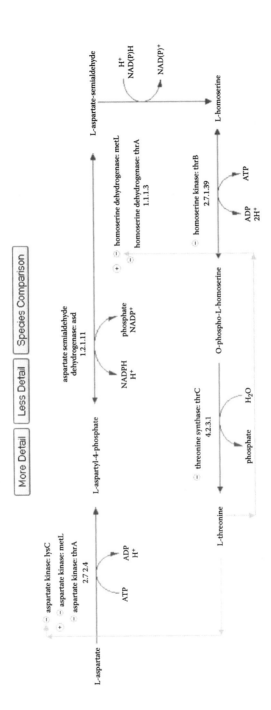

FIGURE 9.18 A screenshot from EcoCyc that illustrates the metabolic pathway for threonine biosynthesis in *E. coli*. Note that the enzymes, gene names, reactants, products, and EC numbers can all be identified, and the relationships between reactions are depicted. (Used with permission from EcoCyc.)

the reaction. The light gray lines illustrate the impact of small molecules on the regulation of enzyme activity, as we discussed in the section entitled "Incorporating Enzyme Inhibitory Effects".

PRACTICE PROBLEM 9.3
PART A

Using Figure 9.18, write out an S matrix for the threonine biosynthesis pathway, beginning with L-homoserine. Ignore the exchange fluxes for now.

SOLUTION TO PART A

The S matrix is:

	v_{thrB}	v_{thrC}
L-homoserine	−1	
ATP	−1	
ADP	+1	
H⁺	+2	
O-phospho-L-homoserine	+1	−1
H₂O		−1
Phosphate		+1
L-threonine		+1

There are two reactions, one of which converts L-homoserine to O-phospho-L-homoserine and another that has L-threonine as the product. Building the S matrix requires including the stoichiometric coefficients for all reactants and products in each reaction. Notice that reversibility does not come into play when building the S matrix; that information is encoded in the constraints, as demonstrated next.

PART B

Set up a linear optimization problem to determine the amount of L-threonine that can be produced from an input feed for L-homoserine of 10 molecules/h. Assume that all required factors are available in the medium except for O-phospho-L-homoserine and write all of the exchange fluxes with positive coefficients in the S_b matrix. How would you code this in MATLAB?

SOLUTION TO PART B

Here, we define an S_b matrix, with all of the exchange fluxes included, and a vector v_b.

$$
\begin{array}{c}
 & v_{thrB}\;\; v_{thrC}\;\; b_{hms}\, b_{ATP}\, b_{ADP}\;\, b_{H^+}\;\; b_{phs}\; b_{H_2O}\; b_{phos}\;\; b_{thr} \\
\begin{array}{c}
\text{hms} \\ \text{ATP} \\ \text{ADP} \\ \text{H}^+ \\ \text{phs} \\ \text{H}_2\text{O} \\ \text{phos} \\ \text{thr}
\end{array}
\left[
\begin{array}{cccccccccc}
-1 & & +1 & & & & & & & \\
-1 & & & +1 & & & & & & \\
+1 & & & & +1 & & & & & \\
+2 & & & & & +1 & & & & \\
+1 & -1 & & & & & +1 & & & \\
 & -1 & & & & & & +1 & & \\
 & +1 & & & & & & & +1 & \\
 & +1 & & & & & & & & +1
\end{array}
\right] = S_b
\end{array}
$$

where hms stands for L-homoserine, phs refers to O-phospho-L-homoserine, phos is the phosphate ion, and thr represents L-threonine, the end product.

$$
\begin{bmatrix}
v_{thrB} \\
v_{thrC} \\
b_{hms} \\
b_{ATP} \\
b_{ADP} \\
b_{H^+} \\
b_{phs} \\
b_{H_2O} \\
b_{phos} \\
b_{thr}
\end{bmatrix} = v_b
$$

With S_b and v_b defined, we can write the problem as follows:

Maximize $-b_{thr}$ (use a negative sign because we want to capture the threonine *leaving* the system)

$$\textbf{Subject to}\quad S_b v_b = 0$$

$$0 \le v_{thrC} \le +\infty$$

$$0 \le b_{hms} \le 10$$

$$0 \le b_{phs} \le 0$$

The units are molecules/h.

Use `linprog` in MATLAB for this problem, as described in the section entitled "Solving FBA Problems Using MATLAB". The text blocks beginning with >> are what I input; the other text is output. First, we define the S_b matrix:

```
>> Sb = [-1 0 1 0 0 0 0 0 0 0;
-1 0 0 1 0 0 0 0 0 0;
1 0 0 0 1 0 0 0 0 0;
2 0 0 0 0 1 0 0 0 0;
1 -1 0 0 0 0 1 0 0 0;
0 -1 0 0 0 0 0 1 0 0;
0 1 0 0 0 0 0 0 1 0;
0 1 0 0 0 0 0 0 0 1]
Sb =
-1 0 1 0 0 0 0 0 0 0
-1 0 0 1 0 0 0 0 0 0
1 0 0 0 1 0 0 0 0 0
2 0 0 0 0 1 0 0 0 0
1 -1 0 0 0 0 1 0 0 0
0 -1 0 0 0 0 0 1 0 0
0 1 0 0 0 0 0 0 1 0
0 1 0 0 0 0 0 0 0 1
```

Next, I define the right side of $S_b v_b = 0$ using the `zeros(m,n)` function, which produces an m-by-n matrix filled with zeros.

```
>> zero_vals = zeros(8,1)
zero_vals =
0
0
0
0
0
0
0
0
```

I can then use the analogous `ones(m,n)` function to begin building my lower- and upper-bound vectors. First, I establish a general lower bound for all of the reactions. In the definition of the problem, most of the lower bounds were negative infinity. For implementation, MATLAB accepts `Inf` as a proxy for an infinite number.

```
>> lower_bounds = ones(10,1)*-Inf;
```

Now, I change some of the lower bounds to zero. First, you can see from Figure 9.18 that the reaction associated with v_{thrC} is irreversible. Next, the problem statement specified that O-phospho-L-homoserine was not present in the medium, so its corresponding exchange flux b_{phs} must also have a lower (and upper) bound of zero. Finally, L-homoserine is the input feed and is not secreted, so b_{hms} must also be positive.

```
>> lower_bounds(2) = 0;        % vthrC
>> lower_bounds(3) = 0;        % bhms
>> lower_bounds(7) = 0;        % bphs
>> lower_bounds
lower_bounds =
-Inf
    0
    0
-Inf
-Inf
-Inf
    0
-Inf
-Inf
-Inf
```

The upper bounds are set similarly, with maximum values of +Inf for most bounds, but with the upper bound of b_{phs} also set to zero as explained above and the upper bound of b_{hms} set to 10 as per the problem statement.

```
>> upper_bounds = ones(10,1)*Inf;
>> upper_bounds(3) = 10;       % bhms
>> upper_bounds(7) = 0;        % bphs
>> upper_bounds
```

```
upper_bounds =
    Inf
    Inf
    10
    Inf
    Inf
    Inf
    0
    Inf
    Inf
    Inf
```

Finally, I write the objective function, which is to maximize the production of L-threonine. In terms of exchange fluxes, we want to move as much L-threonine across the system's boundary, and out of the system, as possible. The exchange flux for L-threonine is b_{thr}, and the positive coefficient for this exchange flux in S_b means that it is moving L-threonine into the system (Figure 9.14 and the corresponding equations). As a result, we want the most negative possible value of b_{thr}; we want to minimize b_{thr}. Conveniently, MATLAB's linprog function always minimizes the objective function, so we can express obj_fxn as:

```
>> obj_fxn = zeros(10,1);
>> obj_fxn(10) = 1;
>> obj_fxn
obj_fxn =
0
0
0
0
0
0
0
0
0
1
```

With all of these steps defined, we can run linprog:

```
>> [vb_vals, obj_val] =
linprog(obj_fxn,[],[],Sb,zero_vals,
lower_bounds,upper_bounds)
```

```
Optimization terminated.
vb_vals =
10.0000
10.0000
10.0000
10.0000
-10.0000
-20.0000
    0
10.0000
-10.0000
-10.0000
obj_val =
-10.0000
```

Remember that the value of the objective ($b_{thr} = -10$) means that 10 molecules of L-threonine can be produced, which requires a positive flux of +10 molecules through the *thrB-* and *thrC*-encoded enzymes. Moreover, we see that 10 molecules/h of L-homoserine, ATP, and H_2O will be required, while 10 molecules/h of ADP and phosphate ion and 20 molecules/h of hydrogen ion will be produced. As we asserted in our definition of the problem, no O-phospho-L-homoserine enters or leaves the system.

APPLICATIONS OF FBA TO LARGE-SCALE METABOLIC MODELS

By organizing reactions into pathways and building pathways into networks, Palsson's team constructed metabolic models of *E. coli* that became increasingly comprehensive. Their latest model (see Orth et al., 2011) accounts for 1,366 genes, which catalyze 2,251 metabolic reactions involving 1,136 unique metabolites.

A model like this can be incredibly powerful, and I'd like to walk you through a few examples. First, similar to what we just did—but on a grander scale—it is possible to predict how much biomass is produced, and which by-products result, when *E. coli* grows in a defined environment. For example, Edwards and Palsson used the first large *E. coli* model (accounting for 600 genes) to determine how the cell used glucose to produce biomass. Figure 9.19a depicts the most central part of the network they reconstructed, which includes glycolysis, the pentose phosphate pathway (PPP), and the TCA cycle.

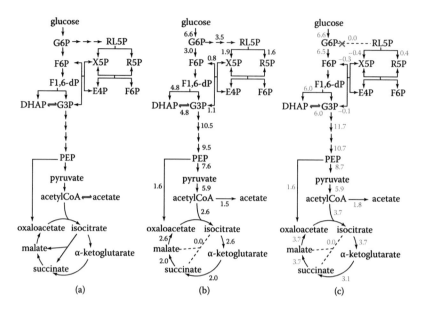

FIGURE 9.19 Modeling flux using FBA. (a) Glycolysis, the PPP, and the TCA cycle pathways; the arrows represent chemical reactions, and the text indicates small molecules. (b) FBA calculation of the flux values in wild-type *E. coli.* Numbers on arrows indicate the fluxes through those pathways. (c) A *zwf⁻* strain in which one of the entry steps to the PPP is knocked out (indicated by the red X), changing the overall flux distribution. Negative numbers indicate that the flux moves in the direction opposite the arrow. (Modified from Edwards, J. S., and Palsson, B. Ø. *Proceedings of the National Academy of Sciences of the United States of America* 2000, **97**(10): 5528–5533. Copyright 2000 National Academy of Sciences, USA.)

Figure 9.19b shows the distribution of fluxes when 6.6 mmol glucose/g DCW/h are available to the cells. Notice how some of the flux goes through the PPP and some goes through the TCA cycle. Also notice how the flux seems to double after glyceraldehyde-3-phosphate because the six-carbon sugar fructose 1,6-diphosphate is split into two three-carbon sugars. Not all of these numbers add up in the figure in the way you would expect; you have to remember that Edwards and Palsson modeled hundreds more reactions than appear here, and some of those reactions pull off flux.

The power of FBA really shines through in Figure 9.19c, where the entry to the PPP from glucose-6-phosphate was (computationally) knocked out. As a result, the PPP runs in the opposite direction, produces more acetate, and has more flux diverted through the TCA cycle.

How would you model these knockouts using FBA? The answer turns out to be extremely simple: For the flux for gene X, v_{geneX}, just add constraints to the upper and lower bounds such that $0 \leq v_{geneX} \leq 0$. Note that knocking out gene X leads to a constraint on protein X. Adding a basic for-loop to this approach would generate the metabolic flux distributions for all single-gene knockouts in a matter of minutes. Two for-loops would get you all of the double-gene knockouts (which has not yet been done experimentally). Add a third for-loop and you can test all of the double knockouts in a variety of environmental conditions. The possibilities are mind-boggling! And it doesn't just have to be knockouts; you can constrain v_{geneX} to nonzero values or ranges. For example, you could reduce the flux for v_{geneX} by 90% from the wild type, similar to a small interfering RNA (siRNA) knockdown. You can also increase the flux from the wild-type value to model the effects of overexpression of gene X.

How well does the *E. coli* FBA model compare to experimental data? The first major comparisons were made by growing *E. coli* on various minimal media, using the measured uptake rates for the carbon source and oxygen as capacity constraints on the FBA model and comparing the resulting growth rate prediction with the experimental measurements. This model (see Edwards, Ibarra, and Palsson, 2001) did very well for certain substrates, such as acetate, succinate, and malate, predicting growth rates that agreed well with experimental measurements.

In other cases, most notably glycerol, the predicted growth rate was much higher than the measured growth rate. Palsson's team hypothesized that if *E. coli* evolved in the glycerol environment, the growth rate would increase and eventually stabilize at the predicted rate. Remarkably, this hypothesis proved to be correct (see Ibarra, Edwards, and Palsson, 2002), one of the most dramatic demonstrations of FBA's ability to predict growth phenotypes.

More commonly, FBA models are used to predict qualitatively whether a given knockout strain of *E. coli* can grow in a given environment. Perhaps not surprisingly, the model performs best with environments that are well studied. The most recent *E. coli* model correctly predicted the essentiality of a given gene in glucose or glycerol minimal medium ~90% of the time (see Orth et al., 2011). With less-familiar environments, the model is less accurate: One study of 110 environments and 125 knockout strains yielded ~65% agreement (see Covert et al., 2004). Nevertheless, these numbers underscore the utility of FBA in predicting growth phenotypes from a small number of measured parameters.

USING FBA FOR METABOLIC ENGINEERING

Some of the most exciting FBA-related work involves the engineering of new *E. coli* strains to make products that had never been made with a biological system. My current favorite of these investigations came from South Korea, where Sang Yup Lee and his colleagues at the Korea Advanced Institute of Science and Technology succeeded in producing the valuable polymer polylactic acid (PLA) directly in cells for the first time (see Jung et al., 2010, and Yang et al., 2010). Among its desirable properties, PLA has low toxicity to humans and is biodegradable, rendering it useful in medical implants (such as screws or scaffolds).

However, PLA production depends on a difficult and expensive two-step synthesis process, so Lee and his team set out to determine whether the entire process could be performed in one step in *E. coli*. Their effort involved challenges in engineering specific proteins, pathways, and networks, so it's a terrific example of metabolic engineering at its finest.

They began with the *E. coli* network with which you are already familiar (Figure 9.20a). Lee and colleagues were particularly interested in lactic acid production, so I have added the chemical reaction that produces lactic acid from pyruvate (catalyzed by lactate dehydrogenase) to Figure 9.20a.

PLA had never been produced biologically, and no enzymes existed for catalyzing the necessary reactions, so Lee and his team had to create new enzymes; they therefore implemented protein-engineering strategies that I can only briefly summarize here. First, they noted that another class of polymers, the polyhydroxyalkanoates, could be synthesized in cells. By introducing several random mutations into the sequences of key genes in the lactic acid pathway, they modified two enzymes such that the pathway would accept lactate as a substrate. They introduced the pathway into *E. coli*: four enzymes total, two of which were optimized (Figure 9.20b).

Although the resulting strain of *E. coli* did produce PLA, it was only a trickle, not nearly the yield needed for economically feasible biological production of PLA. The problem was the trade-off between cellular growth and PLA production; carbon diverted to PLA production cannot be integrated into the necessary biomass components. Moreover, overall yield depends not only on how much PLA each cell can make, but on how many cells are producing it.

Lee's team therefore tried to engineer the organism to increase PLA yield without unduly hindering cell growth. A metabolic engineer typically has two options in this regard: (1) reduce or eliminate the expression

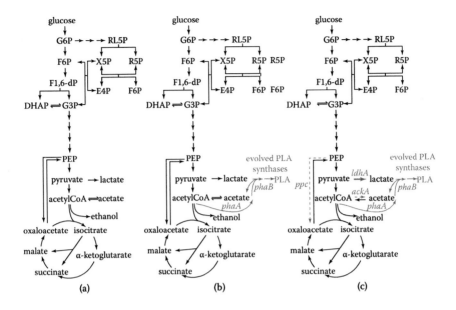

FIGURE 9.20 Harnessing *E. coli*'s endogenous network to generate an "unnatural" product. (a) The glycolytic pathway in *E. coli*, including lactic acid synthesis. (b) Adding four new enzymes to the circuit that use lactate as a substrate. (c) Increasing PLA yield via metabolic engineering. The enzymes encoded by *ackA* and *ppc* were removed from the pathway (dashed lines), and *ldhA* received a stronger promoter (heavy line), thereby increasing the amount of enzyme produced and consequently, the flux. (Modified with permission from Jung, K. Y., Kim, T. Y., Park, S. J., and Lee, S. Y. *Biotechnology and Bioengineering* 2010, **105**(1): 161–171. Copyright © 2009 Wiley Periodicals, Inc.)

of unnecessary genes or (2) introduce or increase the expression of important genes. Lee's team tried both strategies, with the goal of pushing all possible resources to pyruvate and lactate. They removed the gene encoding an enzyme that produces acetate from acetyl-coenzyme A (CoA) (*ackA*), as well as another gene encoding the PEP-carboxylase (*ppc*), which allows flux to bypass pyruvate entirely on its way to the TCA cycle (Figure 9.19c). They also introduced a stronger promoter to increase the flux from pyruvate to lactate (encoded by the lactate dehydrogenase gene, *ldhA*).

The new strain made quite a bit more PLA, but the authors wanted to obtain the greatest possible yield. Their problem was, where do we go from here? There were several options, and they had already made the most obvious modifications. As a result, they turned to FBA to help them design a new strain *in silico*.

As already discussed, it is quite simple to constrain fluxes in FBA, making them higher or lower. Lee's team started with an *E. coli S* matrix, added columns for the metabolic reactions that they had introduced into their strain, and began the design process by considering gene knockouts. They performed simulations in which the PLA synthesis genes were added, and they knocked out each of the genes in the entire metabolic network in turn by setting the upper and lower bounds for the corresponding fluxes to zero. These simulations calculated the resulting growth rate and PLA production rate. Many of the knockouts simply made growth impossible, and many others had no impact on growth or on PLA yield. However, knocking out a handful of genes reduced the growth rate and increased PLA production (Figure 9.21a). The most interesting of these genes (*adhE*) encodes the alcohol dehydrogenase that produces ethanol as part of a two-step pathway starting with acetyl-CoA (shown as a single arrow in Figure 9.21b). The *adhE* knockout was predicted to significantly increase PLA production with only a minor drop in cellular growth rate.

Next, Lee and colleagues used FBA to identify the best candidate genes for overexpression. Once again, there was a trade-off between how fast a cell can grow and the amount of a by-product it can produce because both biomass and by-product require glucose's carbon and energy. Lee and colleagues used FBA to specify this trade-off (Figure 9.22a) by setting the growth rate (both upper and lower bounds) to a specific value as plotted on the x axis in Figure 9.22a. The objective function was then defined as the maximization of PLA production (plotted on the y axis in Figure 9.22a).

The top plot of Figure 9.22a depicts the FBA-calculated trade-off for the wild-type strain. As we would expect, a higher growth rate leads to lower PLA production and vice versa. However, in the right portion of the top plot, a relatively large increase in PLA production results from a relatively small decrease in growth rate. How does gene overexpression affect this trade-off?

Lee and colleagues found that overexpression of different genes could have significantly different impacts on the trade-off curve. For example, increasing the flux through the PPP in FBA (by overexpressing any of the PPP genes) shifts the trade-off down and to the left, leading to both lower growth rates and lower PLA yield (Figure 9.22a, second from top). This result should make intuitive sense to you because drawing flux into the PPP leaves less flux available to lactate and the TCA cycle, from which the

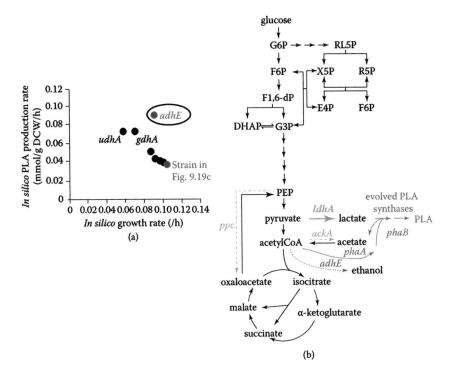

FIGURE 9.21 Simulations reveal a strategy for increasing PLA yield without sacrificing growth rate. (a) Lee and colleagues used simulations to test the results of gene knockouts on PLA production rate and growth rate *in silico*. Knockout of *adhE* was predicted by the boosted PLA production without a substantial loss in growth rate. (b) The protein encoded by *adhE* participates in the ethanol biosynthesis pathway, which is collapsed to a single arrow here. (Modified with permission from Jung, K. Y., Kim, T. Y., Park, S. J., and Lee, S. Y. *Biotechnology and Bioengineering* 2010, **105**(1): 161–71. Copyright © 2009 Wiley Periodicals, Inc.)

protein is produced (a key requirement for biomass production). This was the least-desirable outcome, so the PPP genes were not considered further.

Overexpressing the genes in the TCA cycle pulls the flux past lactate (Figure 9.22a, third from top), generating more cell growth—but at the cost of PLA production. Interestingly, overexpression of the flux through glycolysis actually led to increased growth as well as increased PLA production (Figure 9.22a, second from bottom). Here, the team was limited by engineering concerns; the *E. coli* glycolytic pathway is so tightly regulated that they probably would have had to reconstitute the entire pathway with new regulation, and even then, there would likely have been many problems. They therefore left the glycolytic pathway alone.

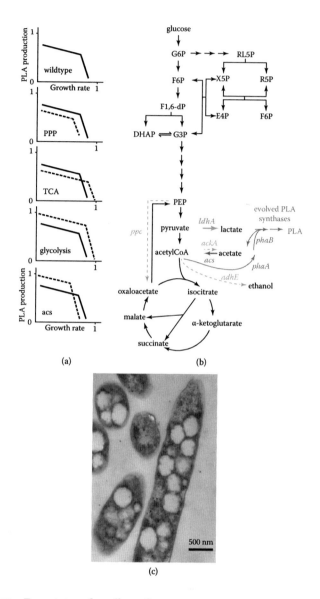

(a)

(b)

(c)

FIGURE 9.22 Examining the effect of gene overexpression on PLA production. (a) Using FBA to shift the bounds of the fluxes and examine the effect on PLA yield. The dashed line depicts the shift in growth versus PLA production when the flux is changed in FBA. (b) Overexpressing *acs* allows the cells to recapture the carbon lost through acetate production. (c) Transmission electron microscopy revealed that the *E. coli* cells harbored large accumulations of PLA (white circles). (Modified with permission from Jung, K. Y., Kim, T. Y., Park, S. J., and Lee, S. Y. *Biotechnology and Bioengineering* 2010, **105**(1): 161–71. Copyright © 2009 Wiley Periodicals, Inc.)

Finally, the team identified a single flux whose overexpression shifted the curve significantly for PLA production, with an acceptable decrease in growth rate (Figure 9.22a, bottom). The corresponding gene (*acs*) encodes acetyl-CoA synthase, which uses acetate as a substrate. They decided to overexpress *acs* in the final strain.

I have to admit that when I first read this paper, even as an expert in *E. coli* metabolism, I was quite surprised. What was going on with *acs*? The result seemed counterintuitive. What I had missed was that acetate was being produced as a by-product of the PLA biosynthesis pathway they had introduced. Overexpressing *acs* enabled the cell to recapture that carbon and push it toward growth (Figure 9.22b).

With these model-derived insights, Lee and colleagues engineered a strain that lacked *adhE* and overexpressed *acs*. They found that the new strain exhibited a 60% increase in polymer content and a 30% increase in lactate fraction relative to their first strain, which meant that the PLA yield was both higher and purer. What I find most compelling is the microscopic image of the bacteria (Figure 9.22c). Those large white spots are the accumulated PLA molecules, and it is pretty clear just by looking at them that Lee's team got about as much PLA out of these bacteria as anyone could expect.

Let's take a step back for some perspective. We've covered two major strategies for examining metabolic networks. One of them delves deep into the kinetic details of individual pathways but is difficult to expand to entire networks; the other is easy to scale up but lacks critical details, notably kinetic parameters, concentrations, regulation, and dynamic information. Both have their advantages and disadvantages. Is it possible to integrate the advantages of each? That question is the main subject of our final chapter.

CHAPTER SUMMARY

Metabolism is a particularly interesting area for systems biology and model building because we benefit not only from a long history of classical methods based on ODE solutions but also from the relatively new application of linear algebra and optimization to enable large-scale modeling of metabolic networks.

One of the best-known mathematical models in biochemistry is the Michaelis-Menten model, which describes the binding of an enzyme to a substrate to form a complex and the subsequent release of enzyme and product. This model begins with four ODEs, which account for the concentrations, production, and loss of substrate, enzyme, complex, and product.

A few assumptions helped us to dramatically reduce the complexity of this system. First, we expanded the terms of the ODEs using the assumption of mass action kinetics, as usual. Next, noting that the equation for E was directly related to the equation for $[C]$ allowed us to establish the conservation of the total enzyme concentration. Thus, we considered only the ODE for $[C]$, realizing that $[E](t) + [C](t) = [E]_0$, the initial enzyme concentration. Further analysis led us to the assumption that $d[S]/dt$ was much smaller than $d[C]/dt$, so for the purposes of our system, we assumed that $[S](t) \approx [S]_0$.

Finally, by considering the system at equilibrium, we identified an equation that described the rate of product formation as a function of the substrate concentration. This rate depended on two parameters: K_m and V_{max}. We demonstrated that although the Michaelis–Menten equation itself is nonlinear, it could be linearized to estimate these two parameter values using a least-squares fit.

The Michaelis–Menten equation reflects only one of the simplest cases in enzyme kinetics, but many other cases can be derived from it, including multiple substrates and a variety of inhibition and regulation scenarios. We mentioned several of these scenarios and derived the equation for competitive inhibition as an example.

In recent years, FBA has been developed for metabolic modeling based on linear optimization. FBA begins with the same set of conservation relationships as the ODE-based approach; however, these relationships are written in terms of fluxes, not concentrations. Considering only the steady-state solutions of these equations leads to a much-studied linear algebra problem in which the product of a known matrix (in our case, containing the stoichiometry of all chemical reactions in the metabolic network) and an unknown vector (containing all of the to-be-determined flux values) is equal to the zero vector. For metabolic networks, the solution of this linear algebra problem is generally a multidimensional space that contains all of the steady-state solutions but none of the dynamic ones.

To estimate the flux values for all reactions in the metabolic network, FBA assumes that the metabolic network has an objective. Most often, this objective is assumed to be the maximization of biomass production (the highest possible growth rate). Biomass production is included in the problem as a pseudoflux that produces all biomass components in their proper ratios.

The FBA linear optimization program is therefore encoded as an objective function with a series of constraints. The first constraint is imposed by the conservation of mass, as described previously. The second set of constraints incorporates the irreversibility of certain chemical reactions, and the third set includes any known maximum reaction rates; most often, this set includes one or more maximum nutrient uptake rates.

With the objective function and constraints defined, FBA returns the predicted values for all fluxes in the metabolic network. This approach has certain limitations, for example: The FBA solution space does not contain dynamic solutions, the output of FBA is in terms of enzyme fluxes (not metabolite concentrations), the FBA solution is generally nonunique, and the assumption of an objective does not always apply. Nevertheless, FBA enables large-scale predictions about metabolic behavior, even with few defined parameters.

RECOMMENDED READING

Chen, W. W., Niepel, M., and Sorger, P. K. Classic and contemporary approaches to modeling biochemical reactions. *Genes and Development* 2010, 24(17): 1861–1875.

Covert, M. W., Knight, E. M., Reed, J. L., Herrgård, M. J., and Palsson, B. Ø. Integrating high-throughput and computational data elucidates bacterial networks. *Nature* 2004, **429**(6987): 92–96.

Edwards, J. S., Ibarra, R. U., and Palsson, B. Ø. In silico predictions of *Escherichia coli* metabolic capabilities are consistent with experimental data. *Nature Biotechnology* 2001, **19**: 125–130.

Edwards, J.S. and Palsson, B. Ø. The *Escherichia coli* MG1655 in silico metabolic genotype: its definition, characteristics, and capabilities. *Proceedings of the National Academy of Sciences of the United States of America* 2000, **97**(10): 5528–5533.

Gunawardena, J. Some lessons about models from Michaelis and Menten. *Molecular Biology of the Cell* 2012, **23**(4): 517–519.

Ibarra, R. U., Edwards, J. S., and Palsson, B. Ø. *Escherichia coli* K-12 undergoes adaptive evolution to achieve in silico predicted optimal growth. *Nature* 2002, **420**(6912): 186–189.

Jung, K. Y., Kim, T. Y., Park, S. J., and Lee, S. Y. Metabolic engineering of *Escherichia coli* for the production of polylactic acid and its copolymers. *Biotechnolology and Bioengineering* 2010, **105**(1): 161–171.

Neidhardt, F. C., Ingraham, J. L., and Schaechter, M. *Physiology of the Bacterial Cell: A Molecular Approach.* Sunderland, MA: Sinauer Associates, 1990.

Orth, J. D., Conrad, T. M., Na, J., Lerman, J. A., Nam, H., Feist, A. M., and Palsson, B. Ø. A comprehensive genome-scale reconstruction of *Escherichia coli* metabolism—2011. *Molecular Systems Biology* 2011, 7: 535.

Overbeek, R. et al. The subsystems approach to genome annotation and its use in the project to annotate 1000 genomes. *Nucleic Acids Research* 2005, 33(17): 5691–5702.

Ro, D. K. et al. Production of the antimalarial drug precursor artemisinic acid in engineered yeast. *Nature* 2006, **440**(7086): 940–943.

Sanghvi, J. C., Regot, S., Carrasco, S., Karr, J. R., Gutschow, M. V., Bolival, B., Jr., and Covert, M. W. Accelerated discovery via a whole-cell model. *Nature Methods* 2013, **10**(12): 1192–1195.

Yang, T. H., Kim, T. W., Kang, H. O., Lee, S. H., Lee, E. J., Lim, S. C., Oh, S. O., Song, A. J., Park, S. J., and Lee, S. Y. Biosynthesis of polylactic acid and its copolymers using evolved propionate CoA transferase and PHA synthase. *Biotechnology and Bioengineering* 2010, **105**(1): 150–160.

PROBLEMS

PROBLEM 9.1
Michaelis–Menten Kinetics: Noncompetitive Inhibition

In Chapter 9, we discussed the effects of competitive inhibition on Michaelis–Menten enzyme kinetics. Now, we consider noncompetitive inhibition, exploring how it impacts the Michaelis–Menten equation and comparing it with competitive inhibition.

We begin by considering noncompetitive inhibition. Here, a substrate S binds an enzyme E to form a complex C1. This complex forms a product P and releases E. Noncompetitive inhibition occurs when an inhibitor I also binds E, but in a different location from S's binding site. When I is bound to E, no P can be made. I can be a part of two complexes (depicted in the figure): enzyme-inhibitor (C2) and enzyme-inhibitor-substrate (C3).

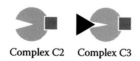

Complex C2   Complex C3

a. Draw the reaction diagram for all reactions in the system. Label the reaction rate constants as follows (noting that k_1, k_{-1}, k_3, and k_{-3} appear more than once because multiple reactions have the same rates):

$$k_1 = \text{rate of forming C1 from E and S}$$

$$k_{-1} = \text{rate of dissociation of C1 into E and S}$$

k_1 = rate of forming C3 from C2 and S

k_{-1} = rate of dissociation of C3 into C2 and S

k_2 = rate of forming P and E from C1 (irreversible)

k_3 = rate of forming C2 from E and I

k_{-3} = rate of dissociation of C2 into E and I

k_3 = rate of forming C3 from C1 and I

k_{-3} = rate of dissociation of complex C3 into C1 and I

b. Write ODEs describing the change over time of all components of the system. Also, write a conservation equation for the total enzyme concentration (equal to the initial concentration of free enzyme, or $[E]_0$), free and bound, in the system.

c. Define the following dissociation constants:

$$K_D = k_{-1}/k_1$$
$$K_I = k_{-3}/k_3$$

Further assume that the formation of complexes C1, C2, and C3 are in equilibrium with respect to the concentrations of S and I. When in equilibrium, a dissociation constant can be written in terms of the substrates and products of the reaction it describes. For example, if the concentrations of two substrates A and B are in equilibrium with their product C:

$$A + B \underset{k_{off}}{\overset{k_{on}}{\rightleftharpoons}} C$$

and

$$K_D = \frac{k_{off}}{k_{on}}$$

then the equilibrium equation is

$$K_D = [A][B]/[C]$$

and

$$[C] = \frac{[A][B]}{K_D}$$

In our system, there are four equilibrium relationships, involving the dissociation constants K_D and K_I. Write two equilibrium equations for K_D and two equilibrium equations for K_I. Simplify your equations so that you have one equation each for [C1], [C2], and [C3]; each equation should depend only on [E], [S], [I], K_D, and K_I.

d. Substitute your equilibrium equations into the equation for $[E]_0$. Then, use your equilibrium equation for [C1] to eliminate [E] from the equation. Solve this expression for [C1].

e. The velocity v of product formation is $d[P]/dt$. Use the information above to write an equation for v in terms of v_{max}, [S], [I], K_D, and K_I. Let $V_{max} = k_2[E]_0$.

f. Use MATLAB to generate a plot of [S] (x axis) versus v (y axis) for varying [I]. Set the following constants:

```
KD = 0.25; %mM

KI = 1; %mM

Vmax = 50; %mmol/min
```

Let $[I] = \{0, 1, 2, 3\}$ mM. Describe and explain the effect of increasing [I].

g. In the section entitled "Incorporating Enzyme Inhibitory Effects", you learned how to model competitive inhibition. Beginning with Figure 9.11 and Equations 9.28–9.33, use equilibrium equations to derive an expression for the flux $v = d[P]/dt$. How does this equation compare to your equation for noncompetitive inhibition in (e)?

h. Generate plots as in (f), but use your equation for competitive inhibition from (g) and compare and contrast the effects of increasing inhibitor concentration for the two types of inhibition.

PROBLEM 9.2
Product Inhibition

We now look at a special case of enzyme inhibition called product inhibition. In this case, the product of the enzymatic reaction itself can serve as an inhibitor of the enzyme, a type of feedback control. Product inhibition is common in metabolic networks and helps to ensure that no more of the product is produced than necessary. One possible mechanism of product inhibition is:

$$E + S \underset{k_{-1}}{\overset{k_1}{\rightleftharpoons}} C1 \xrightarrow{k_2} E + P \underset{k_{-3}}{\overset{k_3}{\rightleftharpoons}} C2$$

a. Write the system of ODEs that describes this system. You should have an equation each for $[E]$, $[S]$, $[C1]$, $[C2]$, and $[P]$. Also, write the conservation equation(s).

b. Now, assume that the kinetic rate constants for E binding to S and P are fast in comparison to the catalytic rate constant ($k_1 \gg k_2$). In this case, we can assume that our complexes quickly reach a thermodynamic equilibrium, such that:

$$[C1] = \frac{[E][S]}{K_S}$$

$$[C2] = \frac{[E][P]}{K_P}$$

where K_S is the equilibrium dissociation constant for E binding to S, and K_P is the equilibrium dissociation constant for E binding to P.

Use these equations and your equation for $[E]_0$ to solve for $[E]$, $[C1]$, and $[C2]$ as a function of $[E]_0$, K_S, K_P, $[S]$, and $[P]$.

c. Now, substitute your answers from (b) into the ODE for $[P]$ so that $d[P]/dt$ is only in terms of $[P]$ and some constants. Simplify your answer to remove at least one of the terms on the right-hand side.

d. How does your result from (c) compare to the equation for $d[P]/dt$ we produced for competitive inhibition? You may substitute $V_{max} = k_2 \cdot [E]_0$ as before. Explain your answer.

e. Now, assume that the unbound product can be irreversibly exported from the cell at a rate proportional to $[P]$ with a rate constant k_{exp}. Modify your equation from (c) to include this term so that your new equation describes the rate of change of P within the cell over time.

f. Use graphical analysis to investigate the dynamics of P within the cell over time. For simplicity, assume that $[E]_0$, $[S]_0$, K_S, K_P, k_2, and k_{exp} are all equal to 1. Separate your equation in (d) into two parts and graph them on the same axes, with $[P]$ on the x axis and the value of each term of the equation on the y axis. Draw this graph by hand; don't use MATLAB!

g. Draw a vector field on the axes in your plot from (f) and circle any fixed points on the graph. Are the fixed points stable or unstable?

PROBLEM 9.3
FBA and the Production of Artemisinin

The partial biosynthesis of the antimalarial drug artemisinin by Jay Keasling's lab at the University of California at Berkeley is a major success story of metabolic engineering (Ro et al., 2006). Ro et al. engineered the yeast's native farnesyl pyrophosphate pathway to produce intermediate compounds that could be converted to artemisinic acid, a precursor of artemisinin, via introduction of an amorphadiene synthase and a novel cytochrome P450 (CYP71). A simplified version of the target pathway is presented below:

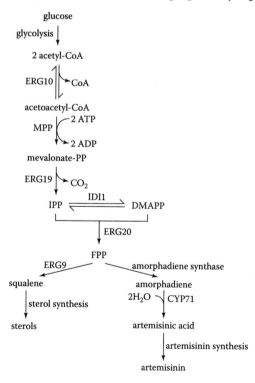

For simplicity, the production of acetyl-CoA from glucose is treated as a single process ("glycolysis"), and several intermediate reactions in the conversion of acetoacetyl-CoA to mevalonate-PP are also treated as a single process. In addition, multiple metabolic steps involved in the synthesis of sterols from squalene have been omitted. Note that glycolysis produces two acetyl-CoA molecules, and ERG10 uses both of these molecules. Similarly, you do not need to consider any of the specific chemical synthesis steps to convert artemisinic acid to artemisinin—they are combined in "artemisinin synthesis".

a. Write the stoichiometric matrix for the synthesis of artemisinin from glucose. You may use a spreadsheet software program such as Microsoft Excel or just include a comment in your MATLAB code.

b. Add exchange fluxes to your matrix for the system input glucose, outputs artemisinin and sterols, and cofactors CO_2, H_2O, CoA, ATP, and ADP. For the purposes of this problem, you will assume that the other, intermediary metabolites (such as mevalonate-PP) are not exchanged.

c. Write code in MATLAB to determine the theoretical maximum amount of artemisinin that can be produced from a maximum glucose import rate of 5 mmol/g DCW/h. Make sure that your irreversibility constraints are consistent with the pathway diagram. Draw or copy the metabolic network on a piece of paper and label the flux values you calculated.

d. The section entitled "Using FBA for Metabolic Engineering" described strategies that were used to maximize the amount of PLA produced by an engineered strain of *E. coli*, one of which was to computationally knock out genes to predict whether each strategy would increase product yield. Perform a gene deletion analysis for all reactions in your network and use a bar graph (MATLAB `bar` function) to plot the production of artemisinin for the wild-type strain and all of the knockouts. Based on your results, which reactions could be knocked out to increase artemisinin production? Justify your answer by labeling the metabolic network with the new set of fluxes determined for one of the knockouts. How does your answer compare with your intuition?

PROBLEM 9.4
FBA and Ethanol Production

Nearly all gasoline at the pump contains ~10% ethanol. Currently, most ethanol comes from crops such as corn and sugarcane, leading to "food-versus-fuel" concerns if food prices rise.

E. coli has a native fermentation pathway for ethanol, but the yield of ethanol is low. Researchers have taken genes encoding enzymes from the ethanol production pathway from *Zymomonas mobilis*, an obligately ethanologic bacterium, and inserted them into E. coli under the control of a common promoter to replace E. coli's native pathway. Mutants with high expression levels of *adh* and *pdc* from Z. mobilis were selected. Please refer to the pathway diagram and equation tables at the end of this problem.

a. Using Microsoft Excel, create an appropriate stoichiometric matrix S for the equation chart. Next, create exchange fluxes to allow for input/output of these metabolites to the system: ADP, H+, phosphate ion, NAD, NADP, CO_2, O_2, and H_2O. Also, create additional fluxes to allow for input of glucose and export of formate and ethanol. You may want to assign all metabolite exchange reactions the same sign in the stoichiometric matrix, so that the sign of each calculated flux indicates transport of that metabolite into or out of the system.

b. Create your lower and upper bounds (*lb* and *ub*, respectively) based on the reversibility of the reactions, as shown in the diagram. Please use +/− Inf as the absolute maximum/minimum, except in the case of glucose import, which is the limiting nutrient in our system. The maximum glucose input is 10 mmol/g DCW/h.

c. Normally, we would assume that our culture is optimizing biomass production. For the purposes of this problem, we define our objective as a simplified "growth" pseudoreaction that requires 2 mmol/g DCW of ribose 5-phosphate (the carbon requirement), 90 mmol/g DCW of ATP (the energy requirement), and 20 mmol/g DCW of NADPH (redox requirement), and produces 2 mmol/g DCW/h of NADH. Create an objective function *f* corresponding to the maximization of flux through the "growth" pseudoreaction. The function should be a 1-by-*n* row vector.

d. In MATLAB, create the matrix S_b and the vectors *lb*, *ub*, and *f* based on your Excel spreadsheet. Remember that S_b does not include *lb*, *ub*, or *f*. A simple implementation is to set up S_b in Excel and read it into MATLAB with `xlsread`. If `xlsread` does not work, save the file as a comma-separated file (.csv) and read it into MATLAB with `csvread`. You can also copy the appropriate cells in Excel, then in MATLAB, Edit → Paste to workspace.

Assume the system is at steady state (S_b v_b = 0) and solve using MATLAB's `linprog` function (find the flux vector v_b). Make a bar chart of the fluxes in the solution (I used `barh`). How much ethanol is produced?

e. You hypothesize that an anaerobic process will increase the yield of ethanol. Test your hypothesis by updating your bounds to represent an anaerobic process. Make a bar chart with the new fluxes. Explain how the pathway fluxes have changed and whether this modification changes ethanol production. How much ethanol is produced? Does the flux going to biomass maintenance change?

f. I hope your result in (e) prompts you to wonder about the trade-off between growth rate and the amount of ethanol produced. Is it possible to squeeze more ethanol out of this network without sacrificing too much growth? Answer this question by plotting the ethanol production flux as a function of the growth flux. Set the bounds of the growth flux to {0, 10%, 20%, 30%, 40%, 50%, 60%, 70%, 80%, 90%, 100%} of the wild-type value you obtained in (d). Then, set the objective to maximize the ethanol production flux. What does your answer suggest about the feasibility of industrial ethanol production with this strain?

Here is the network you will use for Problem 9.4:

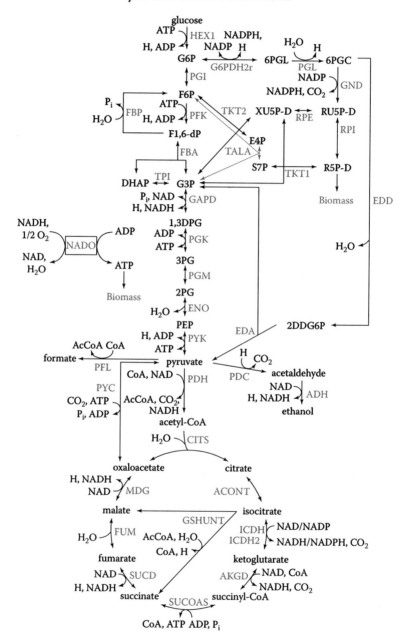

A table of the chemical reactions you will need to include in the matrix follows. The name for each reaction is given at left, the stoichiometry of the reaction appears in the center (with abbreviated metabolite names), and the enzyme name is found at right. Reactions are grouped by pathway. To learn more about these enzymes and metabolites, visit http://www.ecocyc.org and search for "glycolysis."

Glycolysis

ENO	$2PG \rightleftharpoons H_2O + PEP$	Enolase
FBA	$F1,6dP \rightleftharpoons DHAP + G3P$	Fructose-bisphosphate aldolase
FBP	$F1,6dP + H_2O \rightarrow F6P + P_i$	Fructose-bisphosphatase
GAPD	$G3P + NAD + P_i \rightleftharpoons 1,3\ DPG + H + NADH$	Glyceraldehyde-3-phosphate dehydrogenase
HEX1	$ATP + GLC - D \rightarrow ADP + G6P + H$	Hexokinase
PFK	$ATP + F6P \rightarrow ADP + F1,6dP + H$	Phosphofructokinase
PGM	$3PG \rightleftharpoons 2PG$	Phosphoglycerate mutase
PGI	$G6P \rightleftharpoons F6P$	Glucose-6-phosphate isomerase
PGK	$1,3DPG + ADP \rightleftharpoons 3PG + ATP$	Phosphoglycerate kinase
PYK	$ADP + H + PEP \rightarrow ATP + PYR$	Pyruvate kinase
TPI	$DHAP \rightleftharpoons G3P$	Triose-phosphate isomerase

Pentose Phosphate Pathway

EDA	$2DDG6P \rightarrow G3P + PYR$	2-dehydro-3-deoxy-phosphogluconate aldolase
EDD	$6PGC \rightarrow 2DDG6P + H_2O$	6-phosphogluconate dehydratase
G6PDH2r	$G6P + NADP \rightleftharpoons 6PGL + H + NADPH$	Glucose 6-phosphate dehydrogenase
GND	$6PGC + NADP \rightarrow CO_2 + NADPH + RU5P-D$	Phosphogluconate dehydrogenase
PGL	$6PGL + H_2O \rightarrow 6PGC + H$	6-phosphogluconolactonase
RPE	$RU5P-D \rightleftharpoons XU5P-D$	Ribulose 5-phosphate 3-epimerase
RPI	$R5P \rightleftharpoons RU5P-D$	Ribose-5-phosphate isomerase
TALA	$G3P + S7P \rightleftharpoons E4P + F6P$	Transaldolase
TKT1	$R5P + XU5P-D \rightleftharpoons G3P + S7P$	Transketolase
TKT2	$E4P + XU5P-D \rightleftharpoons F6P + G3P$	Transketolase

TCA Cycle

ACONT	$CIT \rightleftharpoons ICIT$	Aconitase
AKGD	$AKG + NAD + CoA \rightarrow CO_2 + NADH + SUCCoA$	Alpha-ketoglutarate dehydrogenase
CITS	$AcCoA + H_2O + OAA \rightleftharpoons CIT + CoA + H$	Citrate synthase

Continued

	TCA Cycle (Continued)	
FUM	$FUM + H_2O \rightleftharpoons MAL$	Fumarase
ICDH	$ICIT + NAD \rightarrow AKG + CO_2 + NADH$	Isocitrate dehydrogenase (NAD+)
ICDH2	$ICIT + NADP \rightarrow AKG + CO_2 + NADPH$	Isocitrate dehydrogenase (NADP+)
MDG	$MAL + NAD \rightleftharpoons H + NADH + OAA$	Malate dehydrogenase
PDH	$PYR + NAD + CoA \rightarrow AcCoA + NADH + CO_2$	Pyruvate dehydrogenase
SUCD	$SUC + NAD \rightleftharpoons FUM + NADH + H$	Succinate dehydrogenase
SUCOAS	$ADP + P_i + SUCCoA \rightleftharpoons ATP + CoA + SUC$	Succinate-CoA ligase

	Anaplerotic Reactions	
GSHUNT	$ICIT + AcCOA + H_2O \rightarrow SUCC + CoA + H + MAL$	Glyoxylate shunt
PYC	$PYR + ATP + CO_2 \rightleftharpoons OAA + ADP + P_i$	Pyruvate carboxykinase

	Fermentation	
ADH	$ACDH + NADH + H \rightleftharpoons EtoH + NAD$	Alcohol dehydrogenase
PDC	$PYR + H \rightarrow ACDH + CO_2$	Pyruvate carboxy lyase
PFL	$PYR + CoA \rightleftharpoons FORM + AcCoA$	Pyruvate formate lyase

	Electron Transport Chain	
NADO	$3ADP + 3P_i + 4H + NADH + \frac{1}{2}O_2 \rightarrow 4H_2O + 3ATP + NAD$	NADH oxidase (ETC)

Next, we have a table of the compounds you will need to include in the matrix. The leftmost column is the metabolite abbreviation, and the next column is the metabolite name. The two rightmost columns are a continuation from the bottom of the leftmost columns, to save space.

1,3DPG	3-phospho-D-glyceroyl phosphate	PYR	Pyruvate
2DDG6P	2-dehydro-3-deoxy-D-gluconate 6-phosphate	R5P-D	Alpha-D-ribose 5-phosphate
2PG	D-glycerate 2-phosphate	RU5P-D	D-ribulose 5-phosphate
3PG	3-phospho-D-glycerate	S7P	Sedoheptulose 7-phosphate
6PGC	6-phospho-D-gluconate	XU5P-D	D-xylulose 5-phosphate
6PGL	6-phospho-D-glucono-1,5-lactone	AcCoA	Acetyl-CoA
ADP	ADP	SUCCoA	Succinyl CoA

Continued

ATP	ATP	SUC	Succinate
CO_2	CO_2	FUM	Fumarate
DHAP	Dihydroxyacetone phosphate	AKG	Alpha-ketoglutarate
E4P	D-erythrose 4-phosphate	CIT	Citrate
F1,6dP	Fructose 1,6-diphosphate	ICIT	Isocitrate
F6P	D-fructose 6-phosphate	MAL	Malate
FDP	D-fructose 1,6-bisphosphate	OAA	Oxaloacetate
FORM	Formate	EtOH	Ethanol
G3P	Glyceraldehyde 3-phosphate	ACDH	Acetaldehyde
G6P	D-glucose 6-phosphate	O_2	Oxygen
GLC-D	D-glucose	CoA	Coenzyme A
H	Hydrogen ion		
H_2O	H_2O		
NAD	Nicotinamide adenine dinucleotide		
NADH	Nicotinamide adenine dinucleotide (reduced)		
NADP	Nicotinamide adenine dinucleotide phosphate		
NADPH	Nicotinamide adenine dinucleotide phosphate (reduced)		
PEP	Phosphoenolpyruvate		
Pi	Phosphate ion		

CHAPTER 10

Integrated Models

- Implement dynamic time-course simulations using FBA

- Integrate transcriptional regulation with metabolic network modeling

- Understand how integrating heterogeneous modeling approaches enables whole-cell models

This book ends with a topic that I've been obsessed with for almost 15 years: the quest to build "whole-cell" models that consider every known gene and molecule. This obsession began just as I started graduate school, when I read an article in *The New York Times* (Wade, 1999) that quoted Clyde Hutchison, who studies *Mycoplasma*, as saying:

> The *ultimate test* of understanding a simple cell, more than being able to build one, would be to build a *computer model of the cell*, because that really requires understanding at a deeper level [emphasis added].

The ultimate test! Hardly a day goes by when I don't think about that quotation; it became my white whale, and I hope by the end of this book that some of you will also want to join the chase.

In 1999, when that article caught my eye, decades' worth of impressive efforts had already gone toward whole-cell modeling. Francis Crick and Sydney Brenner had been talking about "the complete solution of *E. coli*" as part of what they called "Project K" (the "K" stands for the K-12 *E. coli* laboratory strain) as early as 1973 in Cambridge, England. In the 1980s, Michael Shuler at Cornell University made the first effort to model the major biological processes in *E. coli*. At that time, little was known about *E. coli*'s genes, so Shuler built a model based on ODEs that was focused on pathways and processes with coarse-grained representations at first (Domach et al., 1984) and adding detail as the years went by (Shuler, Foley, and Atlas, 2012). Around the same time, Harold Morowitz at Yale University was advocating *Mycoplasma*, the simplest known genus of free-living bacteria, as the best system for a comprehensive mathematical model (Morowitz, 1984). In the 1990s, Masaru Tomita and his colleagues from Keio University in Japan developed a more detailed and gene-based model called E-Cell (Tomita et al., 1999). Of course, Bernhard Palsson was raising metabolic modeling to new heights with the development of FBA, as you learned in Chapter 9 (see Savinell and Palsson, 1992). These and other key efforts all inspired me.

Looking over this impressive body of work, it became clear that no single modeling approach—ODEs, stochastic simulations, or FBA—would be sufficient to model a whole cell. In practical terms, it did not seem possible to use the same approaches to model, for example, the progress of the DNA replication apparatus and the fluxes through the metabolic network. These processes are just too different, in terms of both their underlying physical mechanisms and our ability to describe each of them.

As a result, some people began to focus on bringing different methods together to form integrated models. At the time, this approach seemed to offer some major advantages if it actually worked; you could divide the total functionality of the cell into pieces, model each piece with the best available method, and then put the pieces together into one whole model. Implementing this approach turned out to be a significant advance and led to the construction of the first whole-cell model.

I can now finally reveal my nefarious little plan—to train you all to be whole-cell modelers! Throughout this book, many of the key approaches you'll need to model various functions in a cell were addressed; now, the first critical steps for putting them all together are presented.

DYNAMIC FBA: EXTERNAL VERSUS INTERNAL CONCENTRATIONS

Let's start with using FBA to run dynamic simulations. Wait, didn't we say in Chapter 9 that FBA couldn't be used to simulate dynamics? One of the critical assumptions of FBA is that the metabolic network is at a steady state, right?

Well, yes and no. Imagine that you are growing a culture of *E. coli* in the lab. You have a flask containing glucose-rich medium and some bacteria that are growing exponentially. This system was represented simply in Figure 9.13. After a short period of time, you can assume that the system reaches a steady state, and that the cellular concentrations of metabolites inside the cell will not change appreciably.

However, the concentrations of metabolites *external* to the cell will change, even at steady state. For example, the amount of glucose in the medium will decrease, as will that of other important ions. Other concentrations (for example, the waste products secreted by the cells and even the biomass itself) will increase. These changes in external concentrations are accounted for by using exchange fluxes (Chapter 9, "Steady-State Assumption: Exchange Fluxes").

Figure 10.1 highlights the impact of exchange fluxes on our steady-state assumption. None of the intracellular metabolite concentrations changes over time, but notice that because of the exchange fluxes, the concentrations of N_{out} and X_{out} can change. In other words, FBA cannot be used to simulate dynamics or concentrations *inside the cell*, but due to the exchange fluxes, the dynamic concentration changes *outside the cell* can be simulated.

The strategy for modeling these concentration changes turns out to be similar to building a numerical integrator like the one covered in Chapter 5. In that chapter, we divided the entire simulation into small time steps, then used the values of variables in the previous time step to calculate the values in the next time step via one of several possible methods (Euler, midpoint, Runge–Kutta).

Similarly, for our simulations here, we will divide the time into steps, and then for each step, we will determine the input conditions, calculate the bounds on the FBA problem based on those input conditions, run FBA, and use the FBA results to calculate the output conditions (Figure 10.2).

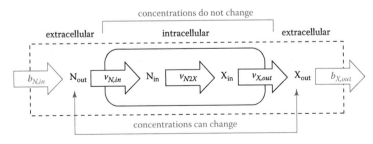

FIGURE 10.1 Modeling changes in concentration over time using FBA. The intracellular metabolites cannot be modeled in this way, but the external concentrations—including the biomass "concentration"—can, thanks to the exchange fluxes.

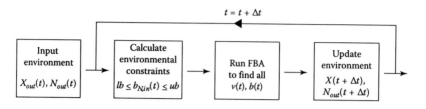

FIGURE 10.2 Schematic of the overall approach for dynamic FBA. Notice that the overall idea is similar to numerical integration: evaluating a function at each time step and using the output of that function to calculate a new value for the variables (see Chapter 5).

ENVIRONMENTAL CONSTRAINTS

Let's apply this strategy to the system in Figure 10.1. Here, our input conditions are the concentrations of nutrient and cells, $[N_{out}](t = t_i)$ and $[X_{out}](t = t_i)$, where i is the current time step. Next, we determine the FBA bounds. You can conceptualize these as environmental constraints; if there are 10 *E. coli* cells and 1 million nearby glucose molecules, the likely metabolic response will be quite different from when only five glucose molecules are present. The constraint will therefore depend on how much substrate is available, how many cells there are, and how long the time step is. We write the equation as follows:

$$b_{N, in}(t = t_i) \le \frac{[N_{out}](t = t_i)}{[X_{out}](t = t_i)\Delta t} \tag{10.1}$$

The numerator is the amount of nutrient at time step i; dividing by the amount of biomass, we obtain the amount of nutrient available to each cell. Finally, dividing this ratio by the time step length yields the maximum amount of nutrient available to each cell over the time interval. That maximum amount is the constraint imposed by the nutrient in the environment.

Notice that $v_{N, in}$ will generally also have a constraint: maximum capacity (Chapter 9, "Defining the Optimization Problem"). Moreover, the environmental constraints are time dependent. This means that when there is plenty of nutrient in the environment relative to the number of cells, the maximum-capacity constraint on $v_{N, in}$ will be more limiting and therefore it will be the dominant constraint. When the nutrient in the environment is largely depleted, then the environmental constraint on $b_{N, in}$ will be more limiting than that of the maximum capacity.

In either case, both constraints are calculated and used as inputs to the FBA problem. FBA then returns the complete metabolic flux distribution for the system, including the values for all of the exchange fluxes. In our example, that would include $b_{N,in}$ and $b_{X,out}$. These flux values are critical to determining the new values of $[N_{out}]$ and $[X_{out}]$.

PRACTICE PROBLEM 10.1

You perform an experiment with E. coli growing exponentially using glucose as the sole carbon source. Initially, you measure a concentration of 10 mmol/L of glucose and 1 g DCW/L in the flask. Several hours later, the glucose concentration has decreased to 0.1 mmol/L, and the dry biomass has increased to 10 g DCW/L.

You decide to simulate the metabolic fluxes at each of these times using FBA, with a time step of 1 min. The maximum uptake rate of glucose for this strain of E. coli was previously measured at 3 mmol glucose/(g DCW · h).

For both time points, determine the environmental constraint imposed by glucose. Which is more constraining, the environment or the maximum capacity constraint?

SOLUTION

At the initial time point t_0, the environmental constraint can be calculated using Equation 10.1 as:

$$b_{N,in}(t=t_0) \leq \frac{[N_{out}](t=t_0)}{[X_{out}](t=t_0) \cdot \Delta t}$$

$$\leq \frac{10 \text{ mmol glucose/L}}{(1 \text{ g DCW/L}) \cdot \text{min}} \qquad (10.2)$$

$$\leq 10 \frac{\text{mmol glucose}}{\text{g DCW} \cdot \text{min}}$$

At the later time point t_1, the environmental constraint will be:

$$b_{N,in}(t=t_1) \leq \frac{[N_{out}](t=t_1)}{[X_{out}](t=t_1) \cdot \Delta t}$$

$$\leq \frac{0.1 \text{ mmol glucose/L}}{(10 \text{ g DCW/L}) \cdot \text{min}} \qquad (10.3)$$

$$\leq 0.01 \frac{\text{mmol glucose}}{\text{g DCW} \cdot \text{min}}$$

The maximum uptake rate is in terms of hours, not minutes, so a conversion is necessary:

$$3\frac{\text{mmol glucose}}{\text{g DCW} \cdot \text{h}} = \frac{3}{60}\frac{\text{mmol glucose}}{\text{g DCW} \cdot \text{min}} = 0.05\frac{\text{mmol glucose}}{\text{g DCW} \cdot \text{min}} \quad (10.4)$$

Comparison of Equations 10.3 and 10.4 shows that when the glucose concentration is high (at t_0), the maximum uptake rate is more constraining; the environmental constraint becomes more limiting as the medium becomes depleted of glucose (at t_1).

INTEGRATION OF FBA SIMULATIONS OVER TIME

With the environmental bounds determined, we are ready to constrain FBA problems. Now it's time to learn how the output of FBA can be used to calculate new external concentrations. Let's begin with calculating the biomass. I mentioned in Chapter 7 that the growth of a population of bacteria in the presence of ample nutrients is often exponential. In particular, it follows this equation:

$$\frac{d[X]}{dt} = \mu[X] \quad (10.5)$$

the solution of which is:

$$[X](t = t_{i+1}) = [X](t = t_i)e^{\mu \Delta t} \quad (10.6)$$

The constant μ is the growth rate, and as you learned in Chapter 9, the biomass exchange flux determined by FBA corresponds to the production of new cells; in fact, it is also equal to the growth rate! So, in this case,

$$\mu = b_{X, out} \quad (10.7)$$

Substituting our system in Figure 10.1 and the growth rate from Equation 10.7 into Equation 10.6 yields:

$$[X_{out}](t = t_{i+1} = t_i + \Delta t) = [X_{out}](t = t_i)e^{b_{X, out} \Delta t} \quad (10.8)$$

Solving Equation 10.8 for $[X_{out}](t = t_{i+1})$ becomes straightforward because $[X_{out}](t = t_i)$ and $b_{X, out}$ are already known, and Δt is chosen by the modeler to be just barely long enough for the steady-state assumption of FBA to hold (a choice of 1–5 s is typical). Notice also that Equation 10.8 can be used for any FBA problem that has a biomass exchange flux; the flux, the time step, and the initial concentration of the biomass are sufficient to calculate the new biomass concentration.

The new value for $[N_{out}]$ can be determined in a similar manner. In this case, it is useful to recognize that the flux is simply the change in concentration over a given time step, normalized by cell biomass. As a result, the ratio of two fluxes at a given time step will be equal to the ratio of the changes in concentration for that same time step. In our example, the ratio of the changes in $[N_{out}]$ and $[X_{out}]$ is equal to the ratio of the corresponding exchange fluxes:

$$\frac{[N_{out}](t=t_i)-[N_{out}](t=t_{i+1})}{[X_{out}](t=t_{i+1})-[X_{out}](t=t_i)} = \frac{b_{N,in}}{b_{X,out}} \tag{10.9}$$

Rearranging, we obtain:

$$[N_{out}](t=t_{i+1})=[N_{out}](t=t_i)-\frac{b_{N,in}}{b_{X,out}}\left([X_{out}](t=t_{i+1})-[X_{out}](t=t_i)\right)$$

$$\tag{10.10}$$

$$=[N_{out}](t=t_i)-\frac{b_{N,in}}{b_{X,out}}[X_{out}](t=t_i)\left(e^{b_{X,out}\Delta t}-1\right)$$

Notice that this formulation depends on the stoichiometric coefficient of the exchange fluxes. In Figure 10.1, we drew the equations such that $b_{N,in}$ is positive when it enters the system, and $b_{X,out}$ is positive when it leaves the system. As a result, Equation 10.10 is a difference: More uptake and growth lead to less available nutrient. Equation 10.10 can also be generalized for all extracellular metabolites that have a corresponding exchange flux in the FBA problem. This includes not only substrates, but also by-products; for a by-product, the second term on the right of Equation 10.10 would be added, not subtracted.

PRACTICE PROBLEM 10.2

Let's go back to the experiment and simulation in Practice Problem 10.1, where you set the bounds for FBA for the initial time point. Remember from Chapter 9 that the biomass pseudoflux (v_{N2X} for this system) converts a number of millimoles of various biomass components into DCW, with units of mmol/g DCW. We assume that 3 mmol of biomass components is converted to 0.01 g DCW.

Given this information, FBA returns:

$$b_{N,in} = 3 \text{ mmol/g DCW/h}$$

$$v_{N,in} = 3 \text{ mmol/g DCW/h}$$

$$v_{N2X} = 0.01/\text{h}$$

$$v_{X,out} = 0.01/\text{h}$$

$$b_{X,out} = 0.01/\text{h}$$

where the overall flux is bounded by the maximum capacity constraint from Practice Problem 10.1.

With this output, 10 mmol/L glucose, and 1 g DCW/L, calculate the glucose and biomass concentrations at $t = 1$ min.

SOLUTION

First, calculate the biomass from Equation 10.8:

$$[X_{out}](t = t_0 + \Delta t) = [X_{out}](t = t_i)e^{b_{X,out}\Delta t}$$

$$[X_{out}](t = t_0 + \tfrac{1}{60}h) = 1 \text{ g DCW}/_L \cdot e^{0.01/h \cdot (1/60)h}$$

$$= 1.0002 \text{ g DCW}/_L$$

Then, calculate the new concentration of nutrient from Equation 10.10:

$$[N_{out}](t = t_{i+1}) = [N_{out}](t = t_i) - \frac{b_{N,in}}{b_{X,out}}\left([X_{out}](t = t_{i+1}) - [X_{out}](t = t_i)\right)$$

$$= 10 \text{ mmol}/_L - \frac{3 \text{ mmol}/_{g\,DCW\cdot h}}{0.01\,\tfrac{1}{h}}(1.0002 - 1)\text{ g DCW}/_L$$

$$= (10 - 0.06)\text{ mmol}/_L = 9.94 \text{ mmol}/_L \text{ glucose}$$

The outputs we calculated in Practice Problem 10.2 would be equal to the inputs we would use for the next time step. As I mentioned earlier, this method is similar to the numerical integration presented in Chapter 5.

The only real difference is in our use of FBA to obtain the exchange fluxes that are used to calculate the next time step. And, as you already learned, FBA can solve thousands of equations simultaneously.

COMPARING DYNAMIC FBA TO EXPERIMENTAL DATA

One of the earliest demonstrations of using dynamic FBA to simulate cell growth, substrate uptake, and by-product secretion is illustrated in Figure 10.3. Palsson and Amit Varma (1994) grew *E. coli* on glucose minimal medium (the environment contains glucose as the sole carbon source

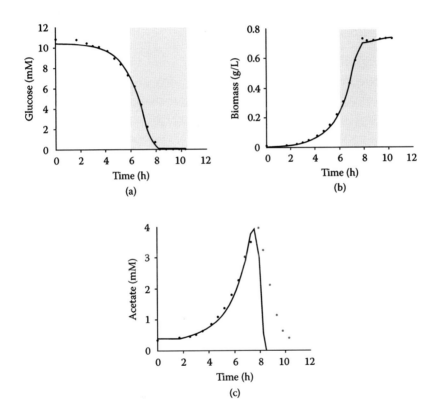

FIGURE 10.3   Dynamic FBA simulation of cell growth in a glucose minimal environment encompasses the changes in glucose concentration (a), cellular biomass (b), and acetate concentration (c). Points represent experimental data. Gray boxes highlight the portion of the simulation for which the environmental constraints have a substantial impact on the simulation. Points that deviate significantly from FBA predictions are highlighted in red. (Modified from Varma, A. and Palsson, B. Ø. *Applied and Environmental Microbiology* 1994, **60**(10): 3724–3731, reproduced/amended with permission from American Society for Microbiology.)

plus a few salts). Under their experimental conditions, *E. coli* produced biomass and the fermentative by-product acetate (for now, look at the data over the range of 0 to 8 h).

These data were also used to simulate their growth conditions. From the glucose data, Varma and Palsson determined the maximum glucose uptake rate; they obtained the maximum acetate transport rates from the acetate data. These two parameters were then used as maximum-capacity constraints in the FBA problem. The overall time course was divided into small time steps (something on the order of seconds or a minute would suffice), and the process in Figure 10.2 was carried out for each time step.

During the first part of the simulation, the glucose concentration was high enough that the environmental conditions were less limiting than the maximum-uptake constraints. This scenario lasted for ~6 h, after which the environmental constraints became significant and increasingly limited the FBA solution (gray in Figure 10.3a,b).

Once the glucose ran out, the cells were left in a pool of acetate that they had secreted, and the cells were able to reuse that acetate (Figure 10.3c). It is important to emphasize that no bounds were changed to produce this simulation; it's simply that the structure of the metabolic network cannot accommodate simultaneous uptake of acetate and glucose—an interesting prediction made by FBA that we also see experimentally! Notice also that the biomass concentration also continues to increase while acetate is being reutilized, but not at the same growth rate (Figure 10.3b).

FBA AND TRANSCRIPTIONAL REGULATION

As mentioned, FBA captures the reutilization of acetate, but not perfectly. The dynamic simulation did not fit the data well (Figure 10.3c) because an extra biological process is at work, one that isn't represented in the FBA framework: transcriptional regulation of gene expression. Acetate reutilization requires the expression of certain metabolic genes that are not expressed in the presence of glucose. Once glucose has been depleted, these enzymes must be synthesized before acetate can be used, and the process of gene expression and protein synthesis takes several minutes.

It is therefore impossible to fit an accurate line to the acetate data without adding information about transcriptional regulation. Unfortunately, FBA as described here has no framework to incorporate such regulation. The seriousness of this problem is underscored by the fact that our acetate example involves only a small number of regulated genes; under typical growth conditions in the lab, such as culture in a nutrient-rich medium, up to ~50% of the genes in *E. coli* can be downregulated!

For my doctoral research, I focused on a way to incorporate transcriptional regulation into FBA. I had already learned about FBA in much the same way you did in Chapter 9, reading about conservation equations, solution spaces, and optimal solutions, and how they could be used to simulate metabolic-network behaviors. My next step was to learn about transcriptional regulation, and I spent about a year and a half in the library, going through the literature on gene expression in *E. coli* so that I could identify what was known and how people represented what they knew. I found that, for the most part, transcriptional regulation was understood at a qualitative level; figures usually showed arrows with a plus or minus attached to them, and manuscripts rarely contained detailed parameters.

As a result, I turned to the Boolean modeling approach described in Chapter 2. This approach seemed like the answer to all of my problems, as it matched the data well and could easily be scaled to represent all known transcriptional regulation in *E. coli*. But how could I integrate it with FBA, which was so different?

TRANSCRIPTIONAL REGULATORY CONSTRAINTS

The critical insight was to realize that regulation imposes constraints. If the expression of a certain gene is downregulated, then the fluxes of the reactions that are catalyzed by the corresponding enzymes must be constrained. In our Boolean representation, that means that both the upper and lower bounds must be set to zero. Like our other constraints, regulation-based constraints can impact the solution space (Figure 10.4) and change the value of the objective function. If our objective in Figure 10.4 is to maximize v_1, then application of the regulatory constraint has a dramatic difference on our computed output.

Regulatory constraints are different from the other constraints we've discussed. They are not only time dependent, like environmental constraints, but also self-imposed: The cell's own machinery imposes the constraint. It might be helpful for you to visualize a large solution space that encloses the full metabolic network available to *E. coli* and then consider a smaller space that is the solution space *at a particular instant in time*, based on the dynamic constraints imposed by the environment and by regulation (Figure 10.4). This smaller space is continually moving around inside, but never outside, the larger space as conditions change.

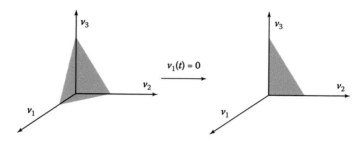

FIGURE 10.4 Regulatory constraints affect the solution space of FBA. If a protein's expression is downregulated at a given time t, then the corresponding enzymatic flux v_1 is set to zero at that time, changing the shape of the solution space. Here, consider the solution space to be a cone that includes positive combinations of fluxes v_1, v_2, and v_3 (left). At some instant in time, an enzyme associated with the flux of v_1 is downregulated. The v_1 flux is therefore constrained to zero, and as a result, the solution space at that point in time only contains solutions with zero values for v_1 (right).

REGULATORY FBA: METHOD

To integrate, we need to incorporate a new step into our dynamic FBA method, one that calculates regulatory constraints (Figure 10.5). When we developed this method, we called it "regulatory FBA" or rFBA for short (Covert, Schilling, and Palsson, 2001).

Let's turn to our glucose uptake/acetate reutilization example for an example. One of the genes required for acetate reutilization that is downregulated when glucose is present is *aceA*, which encodes the isocitrate lyase enzyme AceA. CRP is the transcription factor that induces *aceA* expression. As you know, CRP-based regulation is complicated (Figures 1.1 and 7.1), but for now we'll just say that CRP is inactive when glucose is present.

We can express the information in the preceding paragraph as Boolean statements. First, we write a rule for CRP activity based on the extracellular glucose concentration:

$$\text{CRP}(t) = \text{IF NOT } ([\text{Glucose}](t) > 0)$$

This rule depends only on extracellular glucose concentrations, but you can also write rules based on the activities of other proteins or even on the values of fluxes. Next, we write rules for the expression of *aceA*; we combine transcription and translation into one event called "expression":

$$\text{Expression}_{\text{AceA}}(t) = \text{IF } (\text{CRP}(t))$$
$$\text{AceA}(t) = \text{Expression}_{\text{AceA}}(t) \text{ AFTER SOME TIME}$$

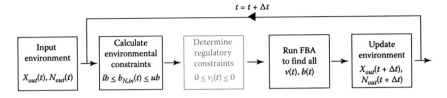

FIGURE 10.5 Schematic of the rFBA approach. rFBA resembles the approach illustrated in **Figure 10.2**, but with a new step (red): the determination of regulatory constraints based on the evaluation of regulatory rules.

These rules would be sufficient for a strictly Boolean model, but because we will be considering time steps, we need to define "AFTER SOME TIME" more carefully. In previous chapters, we talked about how the average gene requires a few minutes to synthesize mRNA and then a few more minutes for translation to produce protein.

Assuming that we want a significant amount of AceA to build up in our system, we could estimate that AFTER SOME TIME means after 24 min (the value that best fit the data in **Figure 10.3**). Then, the previous rule becomes:

$$AceA(t) = Expression_{AceA}(t) \text{ AFTER 24 min}$$

or

$$AceA(t) = Expression_{AceA}(t - 24 \text{ min}) \tag{10.11}$$

In other words, at every time step, the presence of the AceA protein will depend on whether its expression was initiated 24 min before. Notice that, in this model, all of the delays (the "SOME TIME"s) must be specified (24 min). If the rules determine that the protein is absent at time t, then the following constraint will be applied:

$$0 \le v_{AceA}(t) \le 0 \tag{10.12}$$

which is the same as saying that $v_{AceA} = 0$; this is just like modeling a gene knockout, except for the dependency on time. If the protein is present at time t, no regulatory constraint will be applied and the flux v_{AceA} can take on any value. To reemphasize the last sentence: The incorporation of regulatory rules does not mean that the fluxes now only hold Boolean values. The final value of v_{AceA} can hold any real value, including zero. The Boolean rules only specify whether a constraint is applied.

Walking through rFBA (Figure 10.5) is basically the same as for the simulation of dynamics (Figure 10.2). We begin by defining the rFBA stoichiometric matrix, the reversibility and capacity constraints, and the objective function. We write all of the gene expression rules and specify the delay terms. We determine the initial concentrations of all moieties for which an exchange flux has been defined, including cellular biomass. Finally, we choose a time step length and then we're ready to go!

From the initial conditions, you can determine the magnitude of the environmental constraints, as we discussed for dynamic FBA. You can also calculate the regulatory state of the cell (the activities of all transcription factors, whether a given gene is being expressed, and whether the corresponding protein exists) using Boolean modeling. This regulatory state determines the regulatory constraints on metabolism for that time step. Then, you run rFBA, which yields the flux values for all of the internal (v) and exchange (b) fluxes, which you use to update the extracellular concentrations of biomass, substrates, and by-products.

REGULATORY FBA: APPLICATION

The rFBA approach simulates glucose uptake and depletion as well as subsequent reutilization of the acetate by-product (Figure 10.6a,b). Notice that the curve for acetate reutilization now agrees beautifully with the data (gray in Figure 10.6c). Importantly, this model also generates the regulatory network output: For every time point, the activity of all the transcription factors and expression of all of the genes, proteins, and corresponding flux bounds are calculated and stored by the simulation. A few of the most important time points, transcription factors, and genes appear in Figure 10.6d.

Interestingly, even this simple model was able to resolve a mystery that had recently arisen in the literature. At that time, the gene expression microarray had just been invented. This technology enabled the simultaneous measurement of the expression of thousands of genes (Sidebar 1.2), providing the first exciting glimpses of whole transcriptional regulatory networks. James Liao and colleagues at the University of California, Los Angeles, performed one of the earliest studies of this kind in *E. coli*, and they studied growth on glucose and acetate, the same as our example system (Oh and Liao, 2008). Notice that the experimental measurements are a 100% qualitative match with the model's predictions (Figure 10.6d).

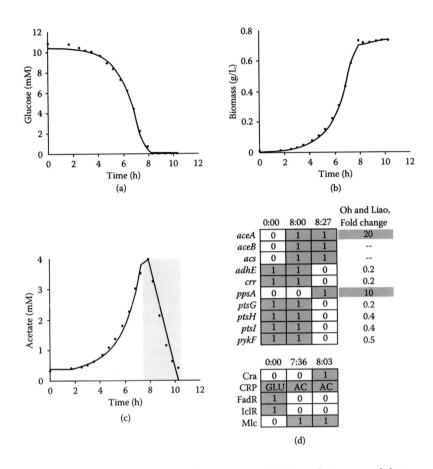

FIGURE 10.6 Adding regulatory information via rFBA results in a match between experimental data and simulated data. (a), (b) Experimental data (points) and simulations (curves) for changes in the glucose concentration in the medium (a) and the cellular biomass in the system (b). (c) rFBA (curve) yields a much better fit to the acetate data (points), particularly during the period of acetate reutilization (gray box). (d) Key features of Oh and Liao's (2008) experimental data compared to model predictions. GLU, glucose; AC, acetate. Time points appear as the column labels (hours:minutes). (Modified from Covert, M. W. and Palsson, B. Ø. *The Journal of Biological Chemistry* 2002, **277**: 28058–28064, reproduced/amended with permission of the American Society for Biochemistry and Molecular Biology.)

Now, for the mystery: Some of the genes seemed to be shifting for no known reason! As Oh and Liao wrote:

> Surprisingly, despite the extensive work on *E. coli* physiology in different carbon sources, *still many genes are regulated for unknown*

purposes by unknown mechanisms. For example, several genes were unexpectedly up-regulated, such as *pps* [and others] in acetate ... or down-regulated, such as [others] and *adhE* in acetate ... [emphasis added]. (Oh and Liao, 2008, page 285)

When I first read this article in detail, I did a double take because I knew that my model was predicting the expression changes of these genes correctly. In other words, the mechanisms were not "unknown," they simply were complex! All I had to do was examine the output of my model to see what was causing rFBA to predict this expression.

The answer lay in an interaction between the metabolic and regulatory networks. CRP switches from OFF to ON in the regulatory network once glucose is depleted, at t = 7 h, 36 min (Figure 10.6d). The switch turns on *aceA*, *aceB*, and *acs* 24 min later ("AFTER SOME TIME"). As a result, the metabolic network switches from glycolysis (glucose utilizing) to gluconeogenesis (glucose synthesizing), allowing the cell to use acetate as a carbon and energy source.

The metabolic switch triggers a shift in the activity of another transcription factor, Cra (catabolite repressor/activator). Cra turns on at 8 h, 3 min (Figure 10.6d), and as a result of Cra activity, *adhE* and other genes are downregulated and *ppsA* is upregulated (Figure 10.6d). The model's capacity to "figure this out" was surprising to *E. coli* experts and extremely exciting to me at the time.

In the end, rFBA enabled us to make several significant advances in modeling *E. coli*. First, we were able to build an integrated regulatory-metabolic model that accounted for 1,010 genes: 906 metabolic genes and 104 transcription factors that regulated over half of those genes (Covert et al., 2004). We showed that this model had better predictive capacity (a greater number of correct predictions) and a broader scope of predictions (increased environmental and genetic perturbations) than using only the metabolic network.

TOWARD WHOLE-CELL MODELING

The development of rFBA also turned out to be a major step toward building whole-cell models, although I did not realize it at the time. The idea to separate the total functionality of the cell into smaller pieces that could be modeled individually and then integrated was a critical aspect of our future strategy for modeling entire cells (Figure 10.7).

The current whole-cell model of the bacterium *Mycoplasma genitalium* includes 28 submodels that are represented using mathematical

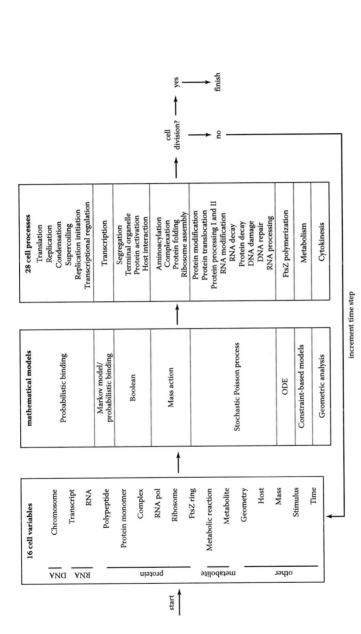

FIGURE 10.7 Schematic of the whole-cell modeling process. The overall approach is the same as in Figures 10.3 and 10.5, but with many more biological processes to model and a very heterogeneous data set that was divided into 16 types of "cell variables." Note that several types of mathematical models were integrated. (Adapted by permission from MacMillan Publishers Ltd.: Gunawardena, J. *Nature Biotechnology* 2012, **30**: 838–840. Copyright 2012.)

modeling approaches that can be roughly characterized into eight groups (Karr et al., 2012). We quickly discovered that the variables also constituted a heterogeneous set that needed to be divided. For example, the number of mRNAs of a particular gene is a very different type of data than the location on the chromosome that binds a certain DNA repair protein. As a result, we grouped cell variables into 16 categories.

I admit that Figure 10.7 looks complicated. Don't let that worry you, because the overall approach is essentially the same as that in Figures 10.3 and 10.5: The variable values are used as input to the submodels, which can be run in series or in parallel, and the outputs of the submodels are then fed to the variables. Again, we choose short (one second) time steps and assume that the submodels operate essentially independently of each other during that time step, an assumption that renders a simulation this complex computationally feasible.

A detailed treatment of the whole-cell model would be the subject of another book (the Karr et al., 2012, article has a supplement available online that is ~120 pages!). However, I would like to include some details of our model to help you tie everything together.

The model in Figure 10.7 is based on *M. genitalium*, a small bacterial parasite that causes a sexually transmitted disease in humans; more relevant to our purposes, it is thought to be the simplest of all culturable bacteria. We selected it as the first cell to model precisely because it is "simple" and "culturable." *M. genitalium* has only 525 genes in its genome, about an eighth of the number of genes in the *E. coli* genome. We also knew that a critical part of our efforts would be to test and validate our model experimentally. We scoured ~900 published investigations and databases to gather ~1,700 known biological parameters. We defined the submodels and variables in Figure 10.7 based on this information.

I have already indicated that breaking up cellular functionality was critical to the model's success, but we implemented a second insight as well: We focused on modeling *one single cell* for *one single cell cycle*. This strategy might at first seem nonintuitive; it certainly did to me. After all, most data on cells come from populations of cells, not single cells!

Modeling a single cell cycle provides several advantages. For one thing, you can keep track of each molecule individually. For example, look at Figure 10.8, a beautiful painting of a cross section of *Mycoplasma mycoides*. This image is rendered to scale, and what you should notice immediately is how crowded everything is. Furthermore, you can leverage

Protein synthesis
1. DNA
2. DNA polymerase
3. single-stranded DNA binding protein
4. RNA polymerase
5. mRNA
6. ribosome
7. tRNA
8. elongation factors Tu and Ts
9. elongation factor G
10. aminoacyl-tRNA synthetases
11. topoisomerases
12a. RecA (DNA repair)
12b. RecBC (DNA repair)
13. chaperonin GroEL
14. proteasome ClpA

Metabolic enzymes
15. glycolytic enzymes
16. pyruvate dehydrogenase complex

Membrane proteins
17. ATP synthase
18. secretory proteins
19. sodium pump
20. zinc transporter
21. ABC transporter
22. magnesium transporter
23. lypoglycan

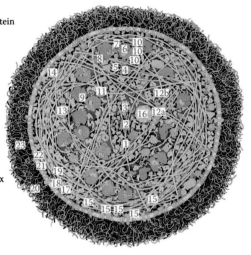

FIGURE 10.8 Painting of *Mycoplasma mycoides*. (This illustration, which is to scale, is used with permission from Dr. David S. Goodsell of the Scripps Research Institute. See http://mgl.scripps.edu/people/goodsell/illustration/mycoplasma.)

your knowledge of the limited space in a cell to make good estimates of molecule numbers. For example, let's examine the ribosomes in Figure 10.8. I count ~20 ribosomes total in the drawing. Considering that this is a cross section through the middle, perhaps with a volume of one-fourth to one-eighth of the total volume, we estimate 80–160 ribosomes in a typical *Mycoplasma* cell (our model uses ~120 ribosomes).

Keeping track of each molecule individually enables several modeling approaches (for example, stochastic simulations). In addition, modeling a single cell imposes strong constraints on the parameter set because the size and mass of the cell at the beginning of the cell cycle must be roughly equal to those of the two daughter cells at the end of the cell cycle.

The whole-cell model has been incredibly exciting for my lab at Stanford University. Our simulations compare well with many different types of data. We have identified emergent properties in the model that help us to better understand how cells work. We have followed up on some of the model's predictions experimentally, and we were thrilled to see that the model correctly predicted biological parameters that had never been measured before (Sanghvi et al., 2013).

We are currently trying to make the whole-cell model widely accessible by making the source code and knowledge bases that we have constructed freely available. Additionally, we're investigating ways to interrogate the terabytes of data that whole-cell modeling has produced. For example, we've developed WholeCellViz (Figure 10.9), a web-based application that enables anyone to explore our simulations in detail (http://wholecellviz. stanford.edu; Karr et al., 2012).

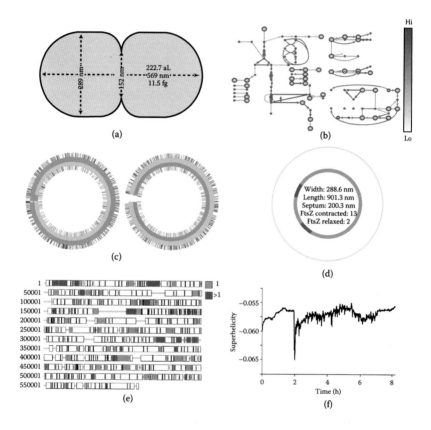

FIGURE 10.9 Visualizing whole-cell modeling data with WholeCellViz. (a) A representation of cell shape. (b) Metabolic map, including fluxes and metabolite concentrations. (c) Map of the *M. genitalium* chromosome, with information about polymerization, methylation, and protein binding. (d) FtsZ contractile ring size. (e) Protein map with information about protein synthesis. (f) Changes in chromosome superhelicity during the cell cycle. (Modified from Lee, R., Karr, J. R., and Covert, M. W. WholeCellViz: data visualization for whole-cell models. *BMC Bioinformatics* 2013, **14**(1): 253. This work is licensed under the Creative Commons Attribution License.)

But, this model is really only the first step into a tantalizing future of whole-cell modeling. Our model needs to be improved, refined, and expanded. Models of more complex cells will need to consider issues such as detailed transcriptional regulation (which is generally absent from *M. genitalium*), compartmentalization (particularly for modeling eukaryotic cells), and detailed spatial modeling using ODEs. Unexpected hurdles will add even more spice to the development of whole-cell models, which is currently a very new (and very exciting!) field of research. The field needs new scientists with in-depth training at the intersection of applied mathematics and molecular biology, which is the major reason I wrote this book. I anticipate that some of you will end up contributing to this effort.

But for now, congratulations!—You have worked with several powerful mathematical, numerical, and analytical techniques, and you've applied them to many circuits, networks, and pathways in cells. I hope you have developed some intuition that will guide you toward determining which methods are useful in particular modeling situations, as well as a feel for how and when to apply them. Most of all, I hope that this knowledge serves you well and that you employ it at the cutting edge of science.

CHAPTER SUMMARY

The last chapter of this book brings together most of the topics that we've covered. First, we discussed how to generate time courses for metabolite uptake and secretion as well as cell growth using FBA (dynamic FBA). This method depends on the exchange fluxes calculated by FBA. The exchange fluxes return the growth rate and uptake and secretion rates, which can be used to determine the changes in biomass and external metabolite concentrations over time. In addition, the environment itself can impose constraints on the FBA model, for example, when the availability of nutrients is limited.

The overall process of dynamic FBA resembles the numerical integrators we discussed in Chapter 5: The value of variables at an initial time point are input into a function that calculates new variable values as an output. These values are then used as the inputs for the next time point, and so on. The FBA function is somewhat more complicated than the functions used for Euler or Runge–Kutta integration, but other than that, the methods are similar.

By comparing dynamic FBA model predictions to experimental data (*E. coli* growing in a flask containing glucose, leading to glucose depletion and secretion and acetate reutilization), we uncovered a critical limitation of FBA: the lack of transcriptional regulatory information. This limitation was addressed by adding regulatory constraints to FBA (rFBA). Regulatory constraints differ from other constraints in that they are both self-imposed and time dependent. The constraints are calculated by evaluating regulatory rules, using the Boolean logic approach developed in Chapter 2. This process occurs at each time step, so rFBA is essentially the same as dynamic FBA, but with the additional step of evaluating regulatory rules.

Finally, rFBA can be seen as a first step toward whole-cell modeling. Like rFBA, whole-cell models bring together multiple modeling approaches into one integrated simulation; like rFBA and dynamic FBA, the process of whole-cell modeling resembles a numerical integrator. However, whole-cell models include many more cellular processes and variables, and although a whole-cell model has been created for *M. genitalium*, the field remains in its infancy.

RECOMMENDED READING

Covert, M. W. Simulating a living cell. *Scientific American* 2014, 310(1): 44–51.

Covert, M. W., Knight, E. M., Reed, J. L., Herrgård, M. J., and Palsson, B. Ø. Integrating high-throughput and computational data elucidates bacterial networks. *Nature* 2004, **429**(6987): 92–96.

Covert, M. W. and Palsson, B. Ø. Transcriptional regulation in constraints-based metabolic models of *Escherichia coli*. *The Journal of Biological Chemistry* 2002, **277**: 28058–28064.

Covert, M. W., Schilling, C. H., and Palsson, B. Ø. Regulation of gene expression in flux balance models of metabolism. *Journal of Theoretical Biology* 2001, **213**(1): 73–88.

Domach, M. M., Leung, S. K., Cahn, R. E., Cocks, G. G., and Shuler, M. L. Computer model for glucose-limited growth of a single cell of *Escherichia coli* B/r-A. *Biotechnology and Bioengineering* 1984, **26**(3): 203–216.

Gunawardena, J. Silicon dreams of cells into symbols. *Nature Biotechnology* 2012, 30: 838–840.

Karr, J. R., Sanghvi, J. C., Macklin, D. N., Gutschow, M. V., Jacobs, J. M., Bolival, B., Jr., Assad-Garcia, N., Glass, J. I., and Covert, M. W. A whole-cell computational model predicts phenotype from genotype. *Cell* 2012, **150**(2): 389–401.

Lee, R., Karr, J. R., and Covert, M. W. WholeCellViz: data visualization for whole-cell models. *BMC Bioinformatics* 2013, **14**(1): 253.

Macklin, D. N., Ruggero, N. A., and Covert, M. W. The future of whole-cell modeling. *Current Opinion in Biotechnology* 2014, 28:111–115.

Morowitz, H. J. The completeness of molecular biology. *Israel Journal of Medical Sciences* 1984, **20**: 750–753.

Oh, M. K. and Liao, J. C. Gene expression profiling by DNA microarrays and metabolic fluxes in *Escherichia coli*. *Biotechnology Progress* 2008, **16**(2): 278–286.

Sanghvi, J. C., Regot, S., Carrasco, S., Karr, J. R., Gutschow, M. V., Bolival, B., Jr., and Covert, M. W. Accelerated discovery via a whole-cell model. *Nature Methods* 2013, **10**(12): 1192–1195.

Savinell, J. M. and Palsson, B. Ø. Network analysis of intermediary metabolism using linear optimization. I. Development of mathematical formalism. *Journal of Theoretical Biology* 1992, **154**(4): 421–454.

Shuler, M. L., Foley, P., and Atlas, J. Modeling a minimal cell. *Methods in Molecular Biology* 2012, **861**: 573–610.

Tomita, M., Hashimoto, K., Takahashi, K., Shimizu, T. S., Matsuzaki, Y., Miyoshi, F., Saito, K., Tanida, S., Yugi, K., Venter, J. C., and Hutchison, C. A., 3rd. E-CELL: software environment for whole-cell simulation. *Bioinformatics* 1999, **15**(1): 72–84.

Varma, A. and Palsson, B. Ø. Stoichiometric flux balance models quantitatively predict growth and metabolic by-product secretion in wild-type *Escherichia coli* W110. *Applied and Environmental Microbiology* 1994, **60**(10): 3724–3731.

Wade, N. Life is pared to basics; complex issues arise. *The New York Times* December 14, 1999.

PROBLEMS

PROBLEM 10.1
Dynamic FBA Concepts

Let's say you want to make a flux balance model for the growth of *E. coli* on ribose. A simplified metabolic network is shown below. Notice that two isomerases, encoded by *rpiA* and *rpiB*, exist in *E. coli* for the interconversion of ribulose 5-phosphate and ribose 5-phosphate. Either of these isomerases can catalyze both the forward and the backward reactions, with fluxes v_{rpiA} and v_{rpiB}, respectively. X represents the cellular biomass (Figure 10.1). Note also the exchange fluxes for ribose, ATP (adenosine triphosphate), ADP (adenosine diphosphate), and X.

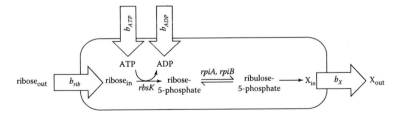

a. Write the flux-balance equations in matrix form. Your answer will include the stoichiometric matrix. Note also that ribose$_{out}$ and X$_{out}$ should not be included in this formulation, as the exchange fluxes are for X$_{in}$ and ribose$_{in}$ (don't worry, they will show up later).

b. Let the initial amount of external ribose be 10 mmol/L, the initial amount of biomass be $(1/e)$ g DCW/L (where e is the base of the natural logarithm), and assume that there is more than enough ATP and ADP available (no environmental constraints caused by ATP or ADP). Transport of ribose into the system is limited such that $0 \leq b_{rib} \leq 1$ mmol/g DCW/min, and the maximum throughput for v_{rpiA} has been measured at 0.5 mmol/g DCW/min. All other bounds are infinite, with the exception of the thermodynamic constraints that are indicated in the diagram. Our objective is to maximize the output of biomass at each time step. Working by hand, calculate the output of the first time step that would be produced using dynamic FBA (including b_{rib}, b_X, [$ribose_{in}$], [X_{out}], and the sum $(v_{rpiA} + v_{rpiB})$). For this problem, assume that 1 mmol of ribulose-5-phosphate is made per 1 g DCW of biomass, so that the units of b_X will be 1/min. Use a time step of 1 min.

c. The expression of *rpiA* is thought to be constitutive; however, expression of *rpiB* occurs only in the absence of the repressor RpiR, which in turn is active only in the absence of external ribose. Assume a time delay of 5 time steps for the amount of protein to reflect changes in transcription. Write the regulatory rules for the system. How would you incorporate this information into your answer to (b)? What changes would you expect to see in your calculations?

d. You would like to generate some double-knockout strains for experiments. Typically, double knockouts are made sequentially; one knockout strain is made first and serves as the platform strain for the second knockout. Which of the following mutants are viable (can produce X) in the presence of glucose as the only carbon source: an *rpiA* knockout; an *rpiB* knockout; an *rpiR* knockout; an *rpiA* knockout followed by an *rpiR* knockout; an *rpiR* knockout followed by an *rpiA* knockout? You may assume that the ribose phosphate isomerase reaction must be catalyzed to produce biomass.

PROBLEM 10.2
Implementing Dynamic FBA

This problem deals with glycolysis, the conversion of one molecule of glucose to two pyruvate molecules through multiple enzymatic conversion steps. The external source of glucose can come from glucose itself or from other molecules such as lactose, one molecule of which can be broken down into a glucose and a galactose molecule by the *lacZ* gene product:

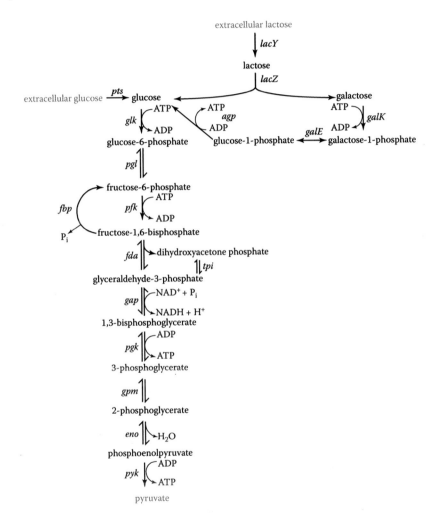

a. Construct a stoichiometric matrix that describes all of the chemical reactions in this system. Let the rows represent metabolites and the columns represent enzymes. It may be easier to create the stoichiometric matrix in Excel and import the data into MATLAB.

b. Add exchange fluxes to your stoichiometric matrix to represent the input/output of extracellular glucose and lactose, ATP, ADP, P_i, NAD^+, $NADH^+$, H^+, and H_2O to and from the system. Assume that the objective of the network is to maximize pyruvate output. Create an exchange flux $b_{pyruvate}$ for this purpose.

c. Load your stoichiometric matrix into MATLAB. If you built your matrix in Excel, you can use the MATLAB function xlsread. Set upper and lower bounds based on the reaction reversibility shown in the diagram. Let the upper bounds through the *pts* and *lacY* transport reactions be 10 mmol/g DCW/h and 5 mmol/g DCW/h, respectively. What does your flux solution look like? Does the cell primarily utilize lactose or glucose catabolism? Graph the value of each flux in v (I used barh for this).

d. We now extend your FBA model of glycolysis to produce a dynamic simulation. Use the following information:

$[G]_0$ = initial concentration of glucose = 10 mmol/L

$[L]_0$ = initial concentration of lactose = 10 mmol/L

$[X]_0$ = initial concentration of dry cellular biomass = 0.1 g DCW/L

Δt = length of the time step = 0.05 h

You may also assume that all other cofactors are present in excess, and that production of 100 mmol of pyruvate corresponds to the production of 1 g DCW of cellular biomass. Plot the concentrations of glucose, lactose, and dry cellular biomass in the medium over 75 h. Describe your observations.

PROBLEM 10.3
Regulatory FBA

Now we will explore the effects of transcriptional regulation on the results of our FBA simulation in Problem 10.2. Recall that the *E. coli lac* operon (Figure 7.1) encodes three genes: *lacZ* (encodes β-galactosidase, which cleaves lactose into glucose and galactose), *lacY* (the permease that transports lactose into the cell), and *lacA* (a transacetylase). The operon is controlled primarily by the *lac* repressor LacI, which prevents transcription of the operon in the absence of lactose, and the glucose-regulated activator CRP. CRP is active in the absence of glucose, and LacI is active in the absence of lactose.

Consider the following regulatory scheme: If the glucose level in the medium is 0.1 mmol/L or greater, CRP is inactive. Once the glucose level falls below 0.1 mmol/L, CRP is activated. In addition, if the lactose level in the medium is 0.1 mmol/L or greater, LacI is inactive. Once the lactose level falls below 0.1 mmol/L, LacI is active. Fifteen minutes after transcription of *lacY* and *lacZ* is initiated, LacY and β-galactosidase are produced (with the same upper and lower bounds as in Problem 10.2).

a. Write the rules for your system and explain qualitatively what you expect to see.

b. Adjust your code from your dynamic FBA in Problem 10.2 to include this transcriptional regulation as a time-dependent constraint on your FBA problem. Note: There will be a period early in the situation (before 15 min of simulation time) in which you will not be able to evaluate the rule for Lac protein expression; in this case, simply use the initial value of *lac* gene transcription at time zero. Plot the activities of CRP and LacI, as well as the transcription and presence of LacY and β-galactosidase, over 75 h. Explain your plots.

c. Plot the concentrations of glucose, lactose, and cellular biomass over time. Compare these plots to those you generated as part of Problem 10.2d. Why do you suppose that the *lac* genes are regulated to produce this behavior?

Glossary

adaptation: In systems biology, the decline in a signal over time.

analytical solving techniques: Mathematical tools that enable the calculation of exact solutions to sets of ordinary differential equations and other conservation relationships.

antibody: An immune-system protein that specifically and reversibly recognizes the three-dimensional structure of a particular molecule. Antibodies are extensively used in biotechnology applications; for example, labeled antibodies against a protein of interest can be used to quantitate the levels of that protein in a sample or to visualize that protein in a cell.

apoptosis: The process in which extracellular and intracellular signals are transduced to bring about programmed cell death.

ATP: Adenosine triphosphate; the molecule that stores energy for metabolism in living cells.

autoregulation: A regulatory motif in which a gene/protein controls its own expression or activity, often in addition to other target genes/proteins.

bi-fan motif: A regulatory motif in which two transcription factors each regulate the same two genes.

bifurcation: In graphical analysis, a change in the number of fixed points in a system.

bifurcation diagram: A plot of a system's bifurcations, which usually result from changing the value of one or more parameters.

bioinformatics: An interdisciplinary field in which biological datasets, often massive in size, are computationally and mathematically analyzed to yield insight into a biological system.

Boltzmann distribution: A distribution that describes a population of particles/molecules in different energy states; specific forms of this distribution raise e to a power that depends on the state energy, the temperature of the system, and a constant.

Boolean: Logic in which an expression can only take two values: true or false (0 or 1, absent or present, etc.).

central dogma: A fundamental paradigm of molecular biology that guided research for many years. The classical formulation of the central dogma is DNA → RNA → Protein. Exceptions to the central dogma, such as RNA → DNA through the enzyme reverse transcriptase, are now recognized.

compartment model: A framework for building sets of ordinary differential equations that represents the values of different variables as compartments, with arrows that indicate the flow from one compartment to another (the terms of the ordinary differential equations).

complex: A chemical species composed of several molecules (with the same identity or different identities) reversibly bound to each other.

concentration: A molecule's abundance in a given volume.

conservation of mass: The principle that all mass in a closed system must remain constant over time.

constitutive: In systems biology, gene expression or protein function that is always "on" (not necessarily at a high level) in the absence of other signals.

continuous: Data that can take any value within a range. Contrast with *discrete*.

control: (1) In systems, the regulation of certain aspects of a system based on other (or even the same) aspects of that system (for example, the regulation of gene circuits via signal transduction and protein activity). (2) In experimental science, an element of an experiment in which the effect of the independent variable is minimized, allowing more rigorous comparison with experiments in which the independent variable is changed.

cooperativity: A synergistic effect in which two or more constituents of a system exhibit a different effect than what would be observed if the constituents acted independently due to interactions among the constituents of the system. For example, in cooperative protein binding, the binding of the first protein molecule to the operator makes it more likely that a second protein molecule will bind nearby.

damping: A process in which the peak of an oscillatory system's behavior becomes lower and lower over each successive cycle of the behavior, eventually suppressing oscillation.

dense overlapping region: A regulatory motif in which a set of transcription factors regulates a common set of target genes.

deterministic: The same initial conditions of a system always lead to the same dynamics and output.

dilution: In systems biology, the loss of concentration of a molecule due to cell division; the number of molecules remains stable, but the volume changes (from one cell to two cells).

dimer: A molecular structure composed of two noncovalently bound subunits, which may be identical (a homodimer) or different (a heterodimer); dimerization is the formation of this structure.

discrete: Data that can only take certain values within a range. Contrast with *continuous*.

DNA-protein interactions: The class of molecular interaction that provides the immediate input to most gene circuits.

dynamics/dynamical: Changes in system components over time.

electrophoretic mobility shift assay (EMSA): A technique based on gel electrophoresis to monitor DNA-protein or RNA-protein interactions. Labeled oligonucleotides are mixed with the protein of interest or a mixture of proteins, then loaded onto a gel and subjected to electrophoresis; oligonucleotides bound to protein migrate through the gel more slowly than free oligonucleotides, as visualized via the label.

elongation: In molecular biology, the addition of subunits to a growing chain; for example, protein elongation is the addition of amino acids to a nascent polypeptide by the ribosome.

Enzyme Commission (EC): A body established by the International Union of Biochemistry to classify and name enzymes and associated molecules and processes.

equilibrium: For a chemical reaction, a state in which the concentrations of reactants and products do not change over time without an external perturbation.

Euclidean distance: The shortest distance between two points, as calculated using the Pythagorean theorem.

eukaryote: A living organism characterized by cells with internal membrane structures such as nuclei.

Euler method: A first-order numerical procedure for solving ordinary differential equations.

exchange flux: In flux balance analysis, fluxes that enable metabolite to enter or leave the system boundaries.

exponent: The number of times that a quantity should be multiplied by itself.

exponential distribution: A probability distribution of a variable in which smaller values of the variable are increasingly more likely to be observed.

extrapolation: The estimation of values beyond the range in which measurements were made.

feed-forward loop: In systems biology, a transcriptional regulatory motif in which expression of a target gene is regulated by two transcription factors, one of which also regulates the expression of the other.

feedback: A control architecture in which system output is used to refine system input or to refine upstream behavior.

first in, first out: A regulatory architecture in which the first gene to be expressed in a system is the first gene to stop being expressed.

first in, last out: A regulatory architecture in which the first gene to be expressed in a system is the last gene to stop being expressed.

fixed points: In graphical analysis, the point at which all of the derivatives of variables in the system are equal to zero.

flux balance analysis (FBA): An approach to modeling metabolic network behavior based on linear optimization.

fusion protein: In biotechnological applications, a single protein composed of the protein of interest (or a portion of that protein) and a reporter (such as green fluorescent protein). Natural fusion proteins consist of protein domains that have been "swapped" during evolution to form new proteins.

gel electrophoresis: A technique in which DNA, RNA, or protein molecules are sorted, usually based on charge or size. The molecules are loaded into a gel, which is placed into a chamber with a salt solution and exposed to an electric current. Over time, the molecules migrate through the gel along the electric field. Special applications of electrophoresis have been developed; for example, two-dimensional protein gel electrophoresis allows a mixed-protein sample, even the entire protein repertoire of a cell/tissue/organism, to be sorted along two dimensions, such as mass and charge.

glycolysis: The metabolic process that breaks down sugars—most notably glucose—to generate energy in the form of ATP and reducing power.

graphical solving techniques: Analyses of sets of ordinary differential equations using a visualization approach. The techniques typically involve generating a graph that illuminates how the equation would behave under a variety of conditions or parameter values.

green fluorescent protein (GFP): A protein, originally from jellyfish, that fluoresces green when exposed to ultraviolet light. Since the 1990s, GFP has been extensively modified into a suite of proteins that are powerful tools for monitoring protein expression and localization.

half-life: The time required for one-half of a substance in a sample to decay or be eliminated; a measure of molecular stability.

housekeeping genes: Genes whose constitutive activity remains approximately stable across many conditions. Housekeeping genes are often used as controls when measuring RNA abundance.

hybrid models: In systems biology, models that combine different modeling approaches together; for example, combining Boolean logic modeling with flux balance analysis.

hybridization: In molecular biology, the process in which one molecule specifically recognizes and binds to another molecule. Most often, the term refers to nucleic acids. Special biotechnological applications of hybridization have been developed. For example, fluorescence *in situ* hybridization enables visualization of the location of a particular DNA or RNA species; a probe is synthesized with a fluorescent label and a sequence that is complementary to the sequence of interest, then introduced to the cell, where it binds (hybridizes) to its complement (the target) and fluoresces.

immunoprecipitation: The process of using an antibody to selectively bind a particular protein from a mixture of molecules. Special biotechnological applications of immunoprecipitation have been developed. For example, chromatin immunoprecipitation is used to investigate DNA-protein interactions; DNA and chromatin are covalently bound to each other with formaldehyde, and then antibodies against chromatin-binding proteins (such as the histones) are used to "pull down" (this term is commonly used, as is the abbreviation IP) the DNA that is bound to these proteins for further analysis.

inflection point: In graphical analysis, the point at which the concavity of a curve changes direction (from concave up to concave down and vice versa).

internally consistent feed-forward loops: A regulatory submotif in which the direct and indirect actions of an upstream transcription factor produce regulation in the same direction (positive or negative); also known as coherent loops.

just in time: An expression pattern in which genes are expressed sequentially rather than simultaneously.

ligand: In biology, a molecule external to the cell to which a receptor binds, often resulting in the activation of a signal transduction cascade.

mass action kinetics: A condition in which the rate of a given chemical reaction is proportional to the product of its reactant concentrations.

mass spectrometry: An experimental technique that provides a profile (or "spectrum") of the relative abundances and relative masses of the molecular constituents of a sample.

metabolism: The biological process in which nutrients are broken down and recombined to produce all of the building blocks needed to make a new cell.

metabolite: An intermediate or product of metabolism, often used to note small molecules in particular.

microarray: A two-dimensional substrate (often smaller than a coin) that enables up to millions of simultaneous measurements of the abundance of molecules present in a sample. A large spectrum of special biotechnological applications of microarrays has been developed. For example, gene microarray analysis is often used to compare the relative amounts of gene expression in two samples, usually with the goal of defining genes whose expression causes the differences between the two samples (such as cancer vs. noncancer, etc.).

model organism: A biological species that is intensively studied to reveal biological phenomena that are difficult to directly study in humans (for example, for ethical reasons). Common model organisms include the bacterium *Escherichia coli*, the yeast *Saccharomyces cerevisiae*, the fruit fly *Drosophila melanogaster*, and the mouse *Mus musculus*.

nonlinear: A mathematical relationship between two variables that does not appear as a straight line when graphed.

northern blotting: A low-throughput method for measuring RNA abundance in a sample by hybridizing a labeled probe against the sequence of interest to a sample.

nuclear localization signal: A domain that allows a protein to be carried across the nuclear membrane.

null space: In mathematics, the solution to the equation $Ax = 0$, where A is a matrix, x is a vector, and 0 denotes the zero vector.

nullclines: In graphical analysis, the curves containing all of the steady-state solutions to an ordinary differential equation in a system.

numerical solving techniques: Mathematical tools that enable the calculation of approximate solutions to mathematical expressions.

objective: In linear optimization, an expression of the variable values that are to be either maximized or minimized.

oligonucleotide: A short, synthetic nucleic acid chain. Oligonucleotides are often labeled and used as probes in biotechnological applications.

operator: A DNA sequence to which a protein can specifically bind to regulate gene expression; also called a *cis*-regulatory element.

operon: In bacteria, a set of DNA sequences that are controlled and transcribed as a single unit and are either translated as a unit or later broken down into individual mRNAs for translation.

optimal solutions: In linear optimization, a set of variable values that maximizes the value of the objective.

order: In numerical integration of ordinary differential equations, the order of a given method is equal to the multiplicative degree of the time step, which is proportional to the error of that method.

ordinary differential equation (ODE): A mathematical expression of the function of an independent variable (such as time) and its derivatives (for example, concentrations of mRNA or protein).

oscillation: Strictly, a behavior that regularly moves between more than one state. For example, systems exhibit oscillation when they transition regularly between active and inactive states or between cellular compartments.

parameter: A measurable factor in a defined system.

parameter estimation: Assigning values to parameters based on empirical data.

proliferate/proliferation: In biology, the process in which a cell grows and divides multiple times to become many cells.

promoter: A DNA sequence upstream of a gene that indicates where transcription should start (the RNA polmerase binding site) and, often, how it should be regulated (operator sites).

pulse-chase labeling: An experimental technique in which a molecule of interest is labeled with a pulse of radioactivity, and its dynamics are followed by monitoring the radioactivity of the sample. For example, proteins can be labeled by adding radioactive amino acids to the medium (the pulse), then adding nonradioactive amino acids to the medium (the chase) before immunoprecipitating the protein of interest.

quantitative polymerase chain reaction (qPCR): A method for simultaneously amplifying and quantitating a DNA or RNA sequence of interest. This technique is often used to confirm the result of high-throughput technologies such as microarrays.

reaction propensity: The probability that a reaction will occur over a given time interval.

receptor: A protein that binds a particular molecule/complex (a ligand) and transduces a signal based on that ligand. For example, in bacteria the chemotaxis receptors are located on the outside of the cell; when the receptor binds a stimulus, the receptor transduces a signal inside the cell to influence whether the cell changes its motion.

reducing power: The storage of electrons by certain compounds, such as reduced-form nicotinamide adenine dinucleotide (NADH) and reduced-form nicotinamide adenine dinucleotide phosphate (NADPH), in order to transfer these electrons via metabolic reactions to produce cellular components.

regulon: A set of target genes that are regulated by the same transcription factor, but their loci may not necessarily be near each other or the gene encoding that transcription factor.

residual: In mathematics, the distance of a point from the line describing the full dataset; the error.

response time: For decay, the time at which the initial protein concentration in a system is reduced by half; for induction, the time at which the protein concentration reaches one-half of the steady-state value.

ribosome: The large and intricate RNA-protein complex that translates messenger RNA molecules into protein molecules.

robustness: The quality of a system that does not change much in response to perturbation (for example, the change in a parameter value).

saddle-node bifurcation: A bifurcation in which two fixed points merge into a single fixed point.

sequencing: The process of determining the order of subunits in DNA, RNA, and protein. Current sequencing technologies are enormously high throughput and require sophisticated computational algorithms for processing and analysis.

signal transduction: The mechanisms by which cells sense and react to factors in the external environment.

single-input module: A regulatory motif in which one transcription factor is solely responsible for the regulation of several genes, often including itself.

solution space: In flux balance analysis, the null space of the stoichiometric matrix, as further constrained by reversibility, maximum capacity, and environmental or regulatory constraints.

spectrophotometry: A technique that measures the absorption of light by a sample. In biotechnical and biochemical applications, spectrophotometry can be used to calculate the concentration of molecules or cells in a sample.

state matrix: A chart that holds the values of all variables in a Boolean system.

steady state: In a set of ordinary differential equations, the condition in which the variables do not further change in value over time.

step response: The reaction of a system to a significant and discrete change from its initial conditions.

stimulus: A molecular signal that elicits a response (for example, from a gene circuit).

stochastic simulation: A computational simulation in which the outputs can change (according to specific probabilities) from run to run; stochastic simulations account for random events.

synthetic circuit: A man-made genetic construct that receives input and generates output. Synthetic circuits are usually modifications of natural biological circuits.

target genes: Genes that are regulated by a particular protein (transcription factor) via signal transduction.

Taylor series approximation: A method, given a certain value and an accompanying function, to estimate a new value of that function based on determining the function's derivatives at the given point.

threshold concentration: In some systems biology applications, the concentration of a protein or other signal that must be present for a consequence to occur (for example, the concentration of active transcription factor required to induce expression).

transcription factor: A protein that regulates the expression of target genes (and often itself) by binding to operator sites for those genes.

transcriptional regulatory network: The set of interactions among genes and proteins that results in particular cellular behaviors.

tricarboxylic acid (TCA) cycle: A metabolic cycle following glycolysis that transfers electrons and oxidizes acetate to yield ATP and other metabolites. Also known as the citric acid cycle and the Krebs cycle.

underdetermined: In mathematics, when there are more variables than equations; there are many possible solutions to underdetermined systems.

uniform distribution: A distribution in which every value is equally likely to be observed.

vector field: In graphical analysis, a representation of how much and in which direction the derivative of one or more variables change at given values of those variables.

western blotting: A low-throughput method for measuring protein abundance in a sample by exposing a labeled antibody against the protein of interest to a sample; also known as immunoblotting.

Index